Metabolic Maps of Pesticides

ECOTOXICOLOGY AND ENVIRONMENTAL QUALITY SERIES

Series Editors: *Frederick Coulston*
and
Freidhelm Korte

OTHER VOLUMES IN THE SERIES

Water Quality: Proceedings of an International Forum
F. Coulston and E. Mrak, editors

Regulatory Aspects of Carcinogenesis and Food Additives: The Delaney Clause
F. Coulston, editor

Environmental Lead
Donald R. Lynam, Lillian G. Piantanida, and Jerome F. Cole, editors

Metabolic Maps of Pesticides
Hiroyasu Aizawa , editor

Metabolic Maps of Pesticides

HIROYASU AIZAWA

Mitsubishi-Kasei Institute of Toxicological
and Environmental Sciences
Yokohama, Japan

1982

ACADEMIC PRESS

A Subsidiary of Harcourt Brace Jovanovich, Publishers

New York London
Paris San Diego San Francisco São Paulo Sydney Tokyo Toronto

ACADEMIC PRESS, INC.
111 Fifth Avenue, New York, New York 10003

United Kingdom Edition published by
ACADEMIC PRESS, INC. (LONDON) LTD.
24/28 Oval Road, London NW1 7DX

Library of Congress Cataloging in Publication Data
Main entry under title:

Metabolic maps of pesticides.

(Ecotoxicology and environmental quality)
Includes bibliographical references and indexes.
1. Pesticides--Metabolism--Handbooks, manuals, etc.
I. Aizawa, Hiroyasu. II. Series.
QP801.P38M47 591.2'1 82-1611
ISBN 0-12-046480-2 AACR2

PRINTED IN THE UNITED STATES OF AMERICA

82 83 84 85 9 8 7 6 5 4 3 2 1

Contents

Foreword xiii

Preface xv

Acknowledgment xvi

1. Acid Amides

Dicryl 2
Kerb (Pronamide, RH-315) 3
MT-101 4
Mebenil (BAS 305F) 5
Barnon (Flamprop-isopropyl) 6
Propachlor 7
Diphenamide 8
Dual (Metolachlor), Antor, Alachlor, and Butachlor (Machete) 9
BAS 3191 11
TTPA 12
Carboxin (Vitavax) 13
Triforine 14
Perfluidone 15

2. Amidines and Guanidines

Galecron (Chlorodimeform) 17
Robenz (Robenidine) 18

3. Anilines and Nitrobenzenes

p-Chloroaniline 20
Propham 21

p-Nitroanisole 22
Chloroanisidine 23
DCNA (Dichloronitroaniline) 24
Dinitramine 25
Dinobuton 26
BBD (*N*-Butyl-4-*tert*-butyl-2,6-dinitroaniline) 27
Trifluralin 28
Basalin 30
PCNB (Pentachloronitrobenzene) 31

4. Biphenyl Ethers

Nitrofen, CNP, Bifenox (MC-4379), Oxyfluorofen (RH-2915), Chloromethoxynyl (X-52), and Preforan 33
Credazine (H-772) 36

5. DDT and Its Analogs

Diphenylmethane and DTE (Diphenyltrichloroethane) 38
DDT [1,1,1-Trichloro-2,2-bis(*p*-chlorophenyl)ethane] 39
DDE [1,1′-(Dichloroethylidene)bis(4-chlorobenzene)] 40
o,p-DDT [1,1,1-Trichlorophenyl-2-(*o*-chlorophenyl)-2-(*p*-chlorophenyl)ethane] 41
DDA [Bis(*p*-chlorophenyl) acetic acid] and DDM [Bis(*p*-chlorophenyl) methane] 42
Chloropropylate 43
Chloromethylchlor 44
Methylchlor 45
Methiochlor 46
Ethoxyaniline 47
Methylethoxychlor 48
Methoxychlor 49
Ethoxychlor 50
Methoxymethiochlor 51
TMMA [N-(α-Trichloromethyl-*p*-methoxybenzyl)-*p*-methoxyaniline] 52

6. Dithio- and Thiolcarbamates

Ethylene bis-dithiocarbamic acid 54
EPTC
(*S*-Ethyl dipropylthiocarbamate)
and Butylate
(*S*-Ethyl diisobutylthiocarbamate) 55
Benthiocarb 56
Molinate (Ordram) 57

7. Five- and Six-Membered Heterocyclic Compounds

Pyrrolnitrin 59
Tachigaren (F-319, Hymexazole) 60
Ipronidazole 61
Isothiazolinones 62
DDOD [3-(3,5-Dichlorophenyl)-
5,5-dimethyl-2,4-oxazo-
lidinedione, Silex] 63
Methazole 64
Oxadiazone 65
Dioxane 66
Terbacil 67
Norfurazone (SAN-9789) and
SAN-6706 68
Dimethirimol 69

8. Imides

DSI [*N*-(3,5-Dichlorophenyl)
succinimide] 71
Procymidon (Sumisclex) 72
Captan 73
Fluoroimide (Spartcide, MK-23) 74

9. Organochlorine Compounds

Lindane 76
Polychlorinated norbornenes 77
Endosulfan 78
cis-, *trans-*Chlordanes 79
Endrin 80
Aldrin 81

HCE (1,2,3,4,9,9-Hexachloro-*exo*-5,6-epoxy-1,4,4a,5,6,7,8,8a-octahydro-1,4-methanonaphthalene) 82
Dihydrochlordenedicarboxylic Acid 83

10. Oxime Carbamates

Thiofanox (DS-15647) 85
Aldicarb 86
Oxamyl 87
Tripate 88

11. Phenoxyacetic Acids

4-CPA (4-Chlorophenoxyacetic acid) 90
2,4-D (2,4-Dichlorophenoxyacetic acid) 91
2,4-DB (2,4-Dichlorophenoxybutyric acid) 92
Phenothiol 93
2,4,5-T (2,4,5-Trichlorophenoxyacetic acid) 94
Dichlorfop-methyl [HOE-23408(OH)] 95

12. Phenyl Ring Fused Five-Membered Heterocyclic Compounds

Fthalide (3,4,5,6-Tetrachlorophthalide) 97
Thiophanate methyl and Thioureidobenzenes 98
Benomyl 99
Parbendazole 101
Methabenzthiazuron 102
Oryzemate 103

13. Phenyl(Aryl) Carbamates

MTMC (Tsumacide) 105
MIPC (Mipcin) 106
BPMC (Bassa) 107
MTBC (*m-tert*-Butyl phenyl *N*-methyl carbamate) 108
Bux 109
Landrin-1 and -2 110
Banol 112
BPBSMC and BPMC (*m-sec*-Butyl phenyl *N*-methyl carbamate) 113
Propoxur (Baygon) 114
MPMC (Meobal) 115
UC-34096 116
Mexacarbate (Zectran) 117
EP-475 and Phenmedipham 118
Carbaryl (NAC, Sevin) 119
Carbofuran (Furadan) 120
Mobam 121

14. Phenylureas and Related Compounds

Siduron 123
Monuron 124
Monolinuron 125
Buturon 126
Clearcide 127
PH-6040 (OMS-1804; Diflubenzuron) 128
BATH (Benzoic acid 2-[2,4,6-trichlorophenyl] hydrazide) 129
Thiadiazuron 130

15. Phosphonothiolates and Phosphonothioates

Leptophos 132
Inezin 133
Dyfonate (Fonofos) 134
N-2596 136

16. Phosphonates

Trichlorfon 138
Surecide 139
Glyphosate (Roundup) 140

17. Phosphorothioamides

Cremart 142
S-2517 143

18. Phosphoramides, Phosphoramidothiolates, and Phosphorimides

Nemacur (BAY-68138) 145
Crufomate 146
Methamidophos 147
Stauffer R-16661 148
Cyclophosphamide 149
Cyolane (Phospholane) 150
Mephosfolan 151

19. Phosphates

GC-6506 153
Dichlorvos 154
Dimethyl and Tetrachlorvinphos 155
Phosphamidon 156

20. Phosphorothiolates

Kitazin P 158
R-3828 159

21. Phosphorothioates

Parathion 161
Sumithion (Fenitrothion) 162
Cyanox 163
Bromophos 164
Dasanit 165
Abate 166
Isoxathion (Karphos) 167

Chlorpyrifos 168
Diazinon 169
Phoxim 170
Salithion 171

22. Phosphorodithiolates

Hinozan (Edifenfos) 173
Mocap 174

23. Phosphorodithioates

Malathion 176
Formothion and Dimethoate 177
Bay NTN 9306 178
Supracide (Methidothion) 179
Phosalone 180
Dioxathion 181

24. Pyrethroids

Pyrethrin I and II 183
Allethrin 184
(+)-*trans*- and (+)-*cis*-Resmethrins 185
cis- and *trans*-Cypermethrins 186
Decamethrin 187
cis- and *trans*-Permethrins 188
Fenvalerate 189

25. Pyridines

CTP (6-Chloro-α-trichloropicoline) 191
DCP (3,6-Dichloro-α-picolic acid) 192
Pyrazon 193

26. Triazines

(*s*-Triazines)
Atrazine 195
Bladex 197

Prometone 198
Cyprazine 198
GS-14254 199
(*As*-Triazines)
Sencor 200
Thiazuril 201

27. Substituted Benzenes and Miscellaneous Compounds

Monochloroacetic acid 203
2,3,6-TBA
(2,3,6-Trichlorobenzoic acid) 204
Dichlorobenil 205
Disugran 206
Chloroneb 207
Bromoxynil 208
PCP (Pentachlorophenol) 209
NK-049 (Methoxyphenone) 210
1-Naphthylacetic acid 211

References 212

Author Index 221

Pesticide Index 228

Subject Index 232

Foreword

Through the study of pesticides and how they affect plants, animals, man, and the environment, many advancements have been made in the evaluation of new and old chemicals. Over the years, various national and international groups have published documents concerning this question of evaluation. Notable among these is the World Health Organization and its monographs, which are the result of deliberations of groups of scientists who are expert in areas relating to the evaluation of chemicals, pesticides, food additives, etc. Many of these monographs are technical reports of the joint F.A.O./W.H.O. meetings, and some are guidelines for evaluation of indirect and direct food additives.

One of the most complicated areas for the understanding of pesticides is the field of metabolic pathways of the chemicals in plants, animals, and man. The author of this book, Dr. Hiroyasu Aizawa, has produced a simple and easy reference that outlines the major pathways of the metabolic fate of the pesticides. As a chemist, Dr. Aizawa has carefully thought out how to present this data and has reduced the enormous amount of information to a simple presentation. This task, in itself, demands high technical skill on the part of the author and allows us to firmly understand which trees exist in the forest. As we move closer to understanding the biological and chemical events that occur when a chemical is introduced into common use, these data become extremely valuable. The many books, pamphlets, and published manuscripts produced yearly make it very difficult to summarize the advances so that a final statement can be made about a single metabolic occurrence.

The Editors of this series find that this book is easily an outstanding contribution to current understanding and further research relating to the biotransformation of pesticides. We feel that this information cannot be found in any other single source, and therefore makes a general contribution to the health of the consumer, to the environment in general and, indeed, to the worker, both in agriculture and in the chemical industry.

Frederick Coulston
Friedhelm Korte

Preface

In the past ten to twenty years, extensive research has been conducted on the metabolism of various pesticides—both for their practical use in controlling destructive pests and for evaluating the effect of their toxicity on the environment. This volume contains a summary of the investigations and drawings of the metabolic patterns on pesticides that were collected with the aid of Chemical Abstracts Service (CAS) for the years 1970 to 1979 [see Menzie (ref. 59a) for references up to 1969]. With the permission of the publishers, some of the maps were taken from the original reports, some were slightly modified for clarity, and some were tentatively drawn from descriptions in the original articles.

The pesticides are classified according to their chemical structures—rather than by the biological activities of insecticides, fungicides, or herbicides—as functional groups or according to common chemical nomenclature. The chemical classifications stress the properties of the mother pesticides and how their degenerative metabolites affect the environment. The systems for decomposing pesticides are shown below the metabolic maps of the pesticides and are summarized later in the book.

As the investigations into pesticide metabolism continue, it has become necessary to compile information as an on-going project. I would appreciate receiving original reports from investigators in this field so that a new compilation could be made on a regular basis.

Hiroyasu Aizawa

Acknowledgments

We gratefully acknowledge permission from the American Chemical Society to reproduce the metabolic maps from the following journals: *J. Agric. Food Chem.* **17**(4)(1969) [138]*; (5)(1969) [164]; **18**(3)(1970) [110]; (3)(1970) [111]; (4)(1970) [13]; (6)(1970) [46,49,112,115]; **19**(1)(1971) [38,42]; (2)(1971) [3,7,165,204]; (3)(1971) [199]; (5)(1971) [145]; (6)(1971) [61,116,170]; **20**(1)(1972) [11,45,50,92,145]; (2)(1972) [109,149,183,184); (4)(1972) [34,47]; (6)(1972) [26]; **21**(2)(1973) [44,48,51]; (3)(1973) [2]; (4)(1973) [43,198,211]; (5)(1973) [18,21]; (6)(1973) [41,90,94,113]; **22**(2)(1974) [24,205]; (3)(1974) [3,112]; (4)(1974) [206]; (5)(1974) [52]; (6)(1974) [68,101]; **23**(2)(1975) [123,140]; (3)(1975) [128,198,205]; (4)(1975) [18]; (6)(1975) [62,129]; **24**(2)(1976) [170]; (3)(1976) [178]; (4)(1976) [181]; (5)(1976) [73]; **25**(1)(1977) [40,78,198]; (2)(1977) [55,80,150,180,290]; (3)(1977) [140]; (4)(1977) [38,42,79]; (5)(1977) [67,151,192]; (6)(1977) [35,187,201]; **26**(2)(1978) [9,130]; (4)(1978) [87]; (5)(1978) [12]. *J. Med. Chem.* **15**(2)(1972) [59].

We also gratefully acknowledge permission from the following journals: *Acta Pharmacol. Toxicol* **30**(1971) [203]; **35**(5)(1971) [36]; (6)(1971) [174]; **36**(2)(1972) [158]; **37**(4)(1973) [103]; (7)(1973) [158,174]; (9)(1973) [63];(12)(1973) [63]; **38**(5)(1974) [105]; (7)(1974) [71]; (10)(1974) [171]; **39**(11)(1975) [167]; **40**(4)(1976) [34]; **41**(10)(1975) [36]. *Arch. Environ. Contam. Toxicol.* **2**(2)(1974) [82]. *Biochem. J.* **96**(1965) [76,100]. *Biochem. Pharmacol* **26**(1977) [66,173]. *Bull. Environ. Contam. Toxicol* **7**(2/3)(1972) [76]. *Chemosphere* **3**(1974) [125]; **5**(1974) [83]. *J.Econ.Entomol.* **66**(1)(1973) [159]. *J. Environ. Sci. Health* **B13**(4)(1977) [146]. *Nature* 237(1972) [77]. *Pestic.Biochem.Physiol.* **1**(1972)[181]; **2**(1972) [148,166]; **7**(1977) [6,126,147]. *Pesticic. Sci.* **2**(1971) [4,23,35,72]. *Phytopathology* **64**(1974) [60]. *Xenobiotica* **1**(6)(1971) [155]; **2**(2)(1972) [154].

*Numbers in brackets [ ] indicate pages on which metabolic maps appear in this volume.

1

Acid Amides

DICRYL

Maximum conversion of dicryl is achieved when fungal growth is ended. Within one week, 22% of the dicryl is transformed to its metabolites, but 34% remains unchanged.

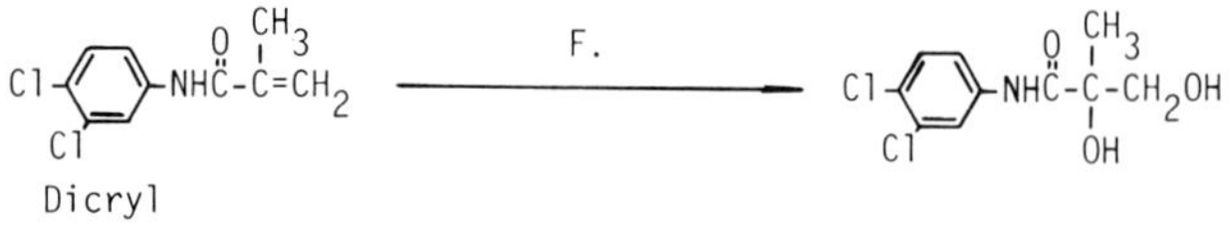

Fungi *Rhizopus japonicus*[1]

KERB (PRONAMIDE, RH-315)

Over the period of the investigation, the final products in the metabolic chain were the metabolites I and II in soil, alfalfa, and rat feces. In rat and cow urine, metabolite I was ultimately converted to metabolite II. All of the identified metabolites were the result of manipulation on the two terminal carbon atoms. Although 3,5-dichlorobenzoic acid could not be identified, dealkylation should occur to form the benzoic acid analog in the presence of unidentified metabolites. Carbonyl-labeled kerb and 3,5-dichlorobenzoic acid, labeled at both the carbonyl and phenyl rings, were metabolized to release the significant levels of CO_2 in nonsterilized soil.

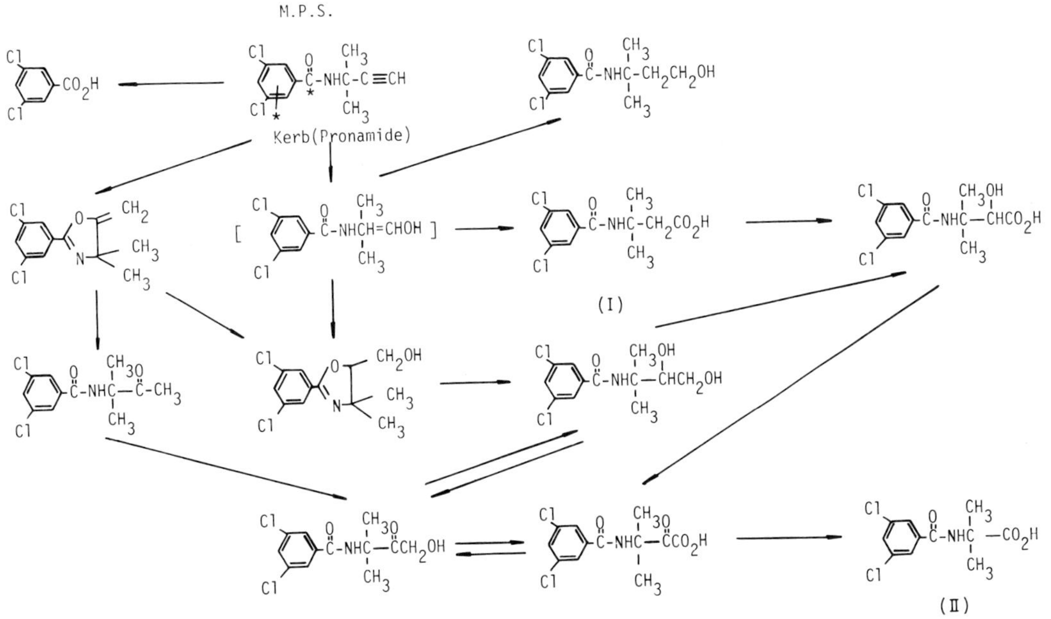

Plant	Alfalfa[2]
Soil	Loam soil[2] Hagerstown soil[3]
Mammal	Rat, Cow[4]

MT-101

The amide bond of MT-101 is hydrolyzed to form its corresponding carboxylic acid and aniline, which was not detected, however. α-Naphthoxyacetic acid is then transformed to several metabolites and 2-hydroxy-1,4-naphthoquinone.

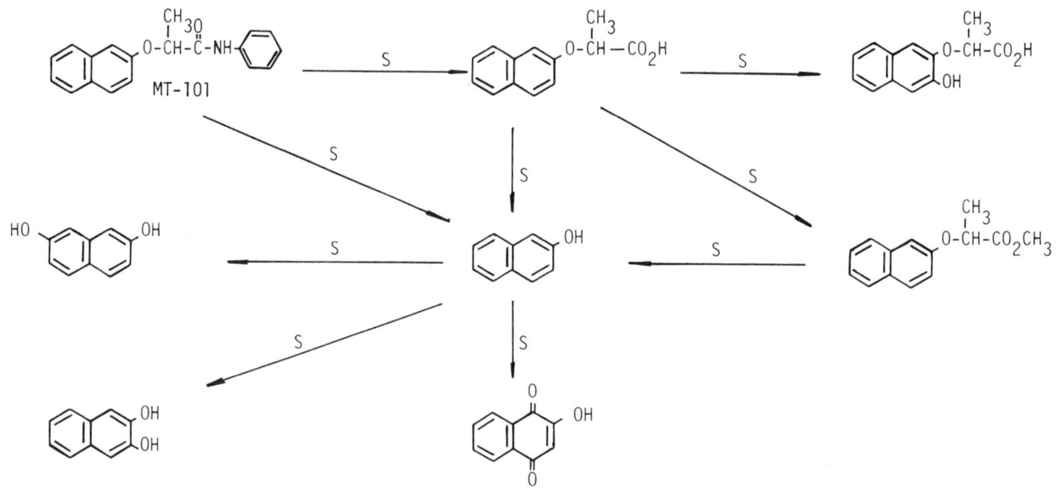

Soil 5

MEBENIL (BAS 305F)

Although aniline has not been detected in the rat liver and kidney homogenate system, one of the metabolic paths must be hydrolysis of the amide bond.

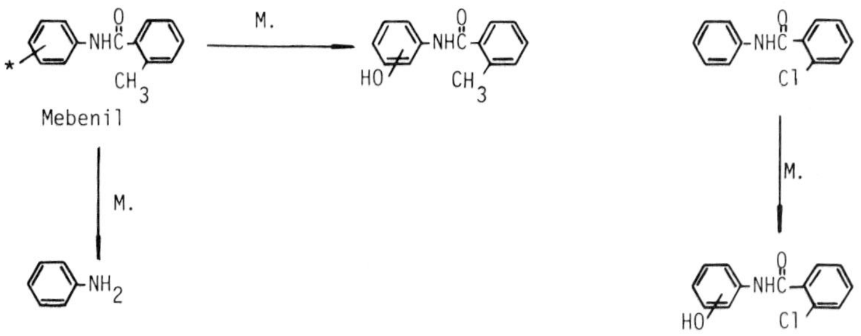

Mammal Female Wistar rats; female rabbit (New Zealand white strain); female guinea pig (Dunkin–Harvey strain category 2)[6]

BARNON (FLAMPROP-ISOPROPYL)

Hydrolysis of ester and amide linkages, cleavage of the C–N bond, and hydroxylation of the phenyl ring (not substituted with halogens) have been observed in the metabolism of barnon in plants.

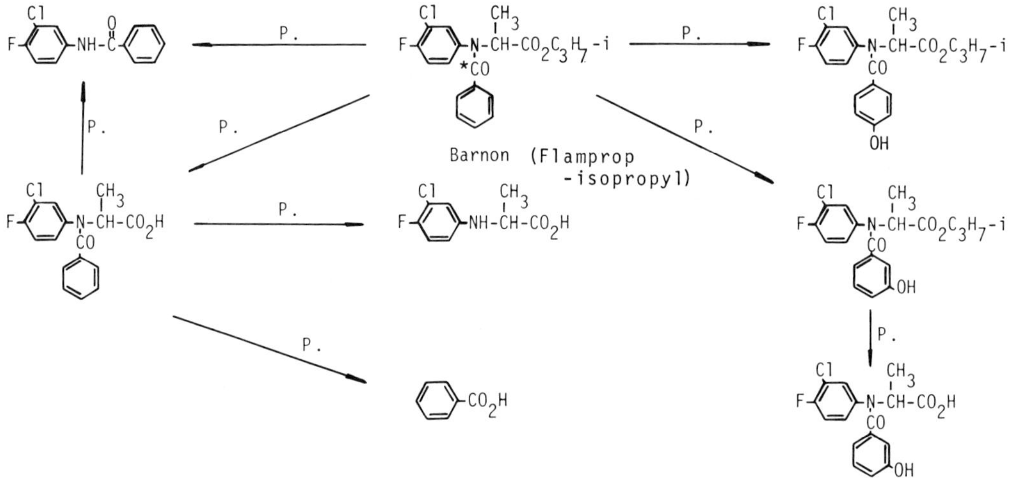

Plant Spring barley (cv. Julia)[7]

PROPACHLOR

The primary mode of metabolism is nucleophilic displacement of the α-chlorine atom of propachlor by a sulfhydryl group of a peptide, e.g., the glutathione or γ-glutamylcysteine conjugate in plants. This displacement proceeds nonenzymatically *in vitro* and both enzymatically and/or nonenzymatically *in vivo*. These metabolites are transient intermediates, not the end products, in plants.

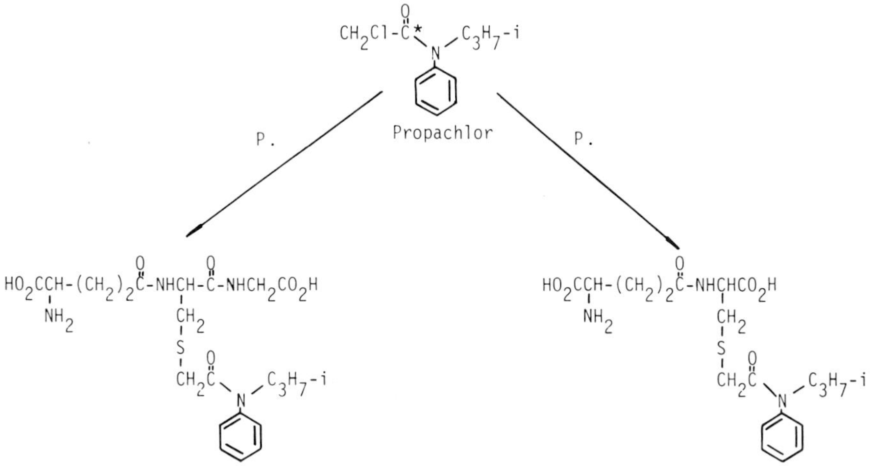

Plant Corn: *Zea mays* L.,var.N.D.KE.47101; Barley: *Hordeum vulgare* L.,var.Larker; Sorghum: *Sorghum vulgare* Pers., var.N.D.104; Sugarcane: *Saccharum officianarum* hybrid C.P.61-37x C.P.56-59[8]

DIPHENAMIDE

Monomethyldiphenamide is the most persistent metabolite of diphenamide; it is further demethylated and hydrolyzed to diphenylacetic acid in plants. Toxicity of the metabolites to the plant is presented in the References.

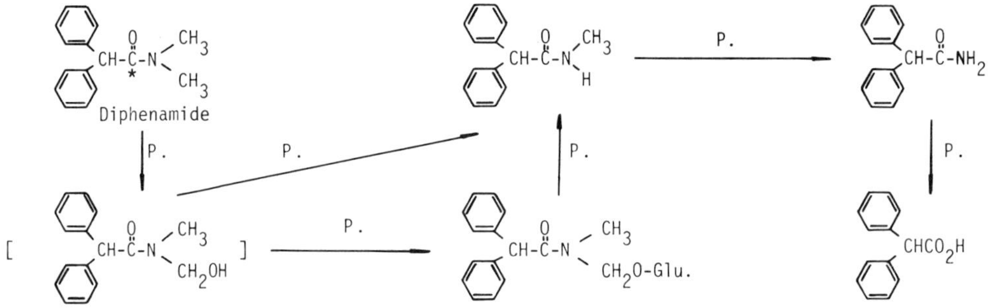

Plant	Well-branched, winged euonymus: *Euonymus alatus* (Thunb); Tomato: *Lycopersicon esculentum* Mill; Bermuda grass: *Cynodon dactylon*[9]
	Wheat: *Triticum aestivum* (L.); Tomato: *Lycopersicon esculentum* Mill.[10]
	Soybean: *Glycine max.* L., Harosoy 63[11]
	Parrotfeather (aquatic plant): *Myriophyllum brasilience* L.; Water hyacinth: *Eichornis crossiper* L.; Waterthread: *Potamogeton diversifolius* Raf.; Algae: *Ulothrix* sp., *Oedogonium* sp., *Gloeocystis* sp., *Scenedesmus* sp.[12]

DUAL (METOLACHLOR), ANTOR, ALACHLOR, AND BUTACHLOR (MACHETE)

In fungi, α-chlorine atoms of these pesticides are not displaced by sulfhydryl groups but by hydroxyl groups although the chlorine atom of propachlor is displaced by a sulfhydryl group in the first degradation step in plants. Dealkylation, deacylation, and ring formation to form indolines and quinolines occur in fungi.

Dual

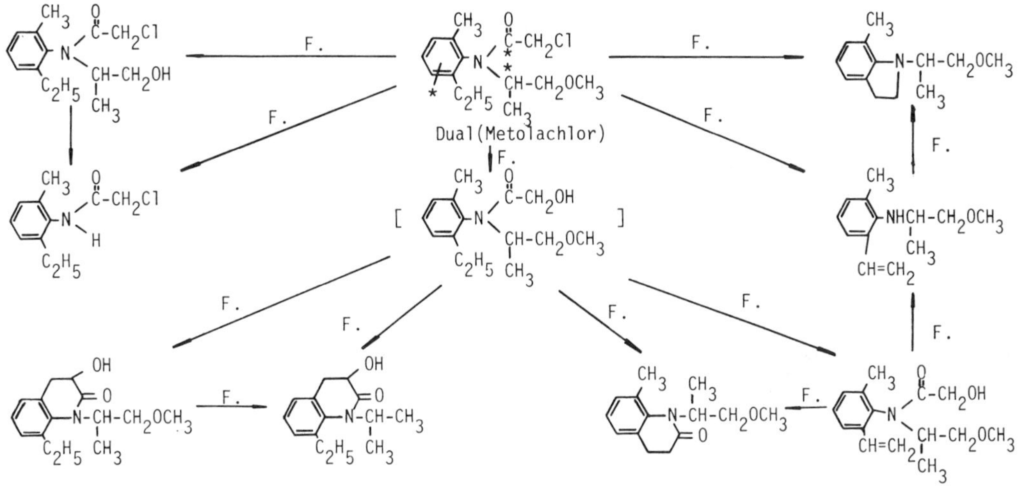

Antor

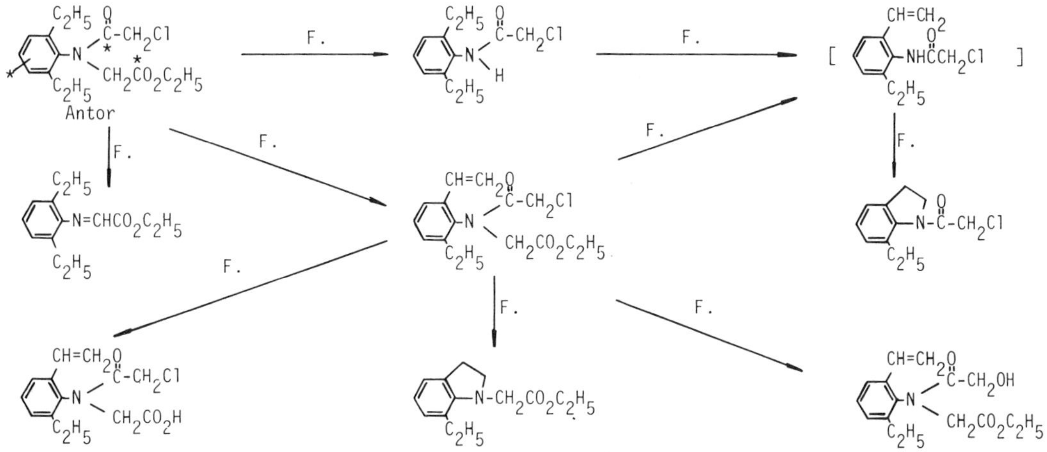

Fungi *Chaetomium globosum*[13]

Alachlor, Butachlor

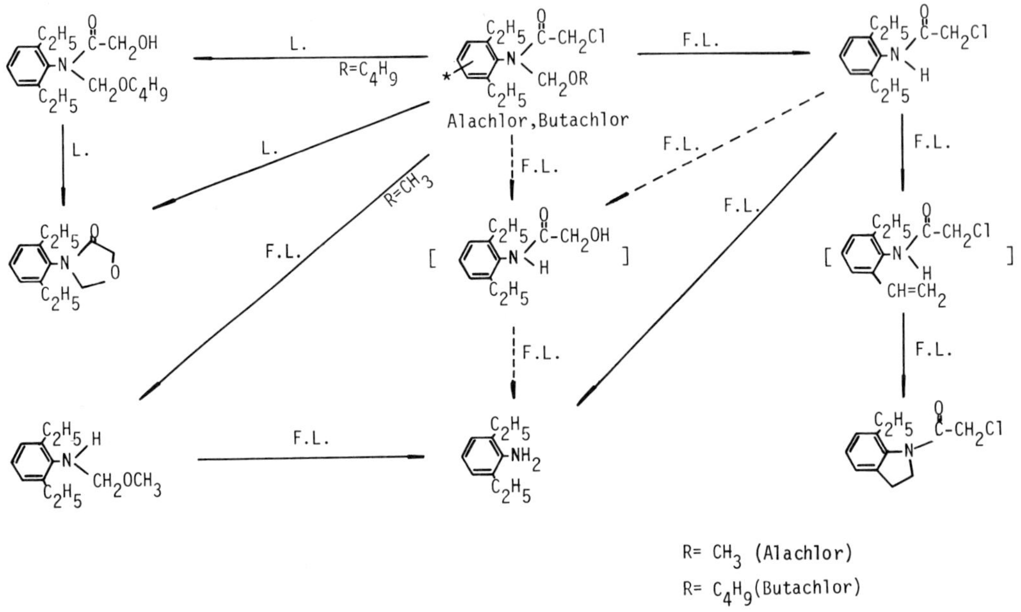

Fungi	Soil fungi: *Chaetomium globosum, C. bostrychodes, Fuzarium roseum;* species of *Penicillium, Phoma, Alternalia, Paecilomyces,* and *Trichoderma*[14]
Light	UV lamp[15]

BAS 3191

Hydroxylation of either of the α-methyl groups of BAS 3191 occurs and these metabolites are not further degraded by fungi.

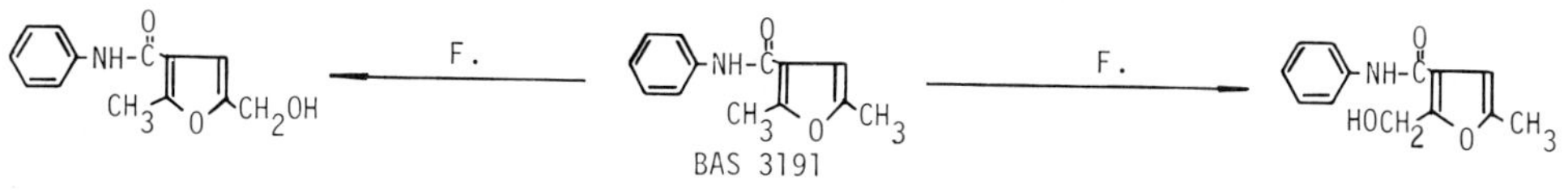

Fungi *Rhizopus japonicus, R. nigricans, R. peka;* two strains of *Mucor*[16]

TTPA

TTPA is metabolized to its corresponding propionic acid without a bromine atom in the 5-position of the pyrazole ring by N-dealkylation, hydrolysis, and position-selective debromination in mammals.

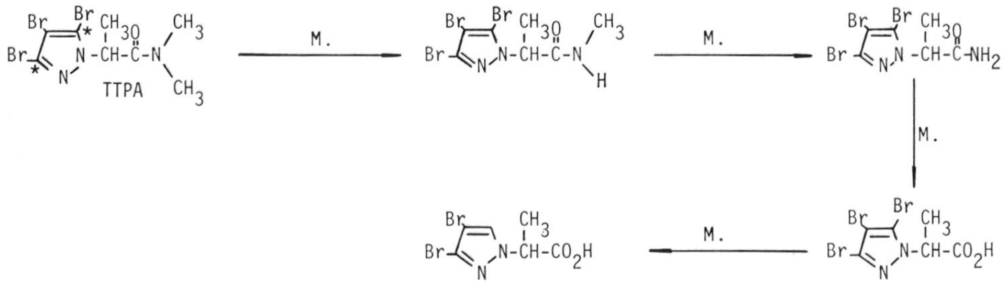

Mammal Female and male S-D rats[17]

CARBOXIN (VITAVAX)

Carboxin forms the anilide-lignin complex as the plant grows. The anilide moiety of the lignin complex may be mainly in the sulfoxide form rather than carboxin itself because of the rapid oxidation of carboxin to the sulfoxide in plants.

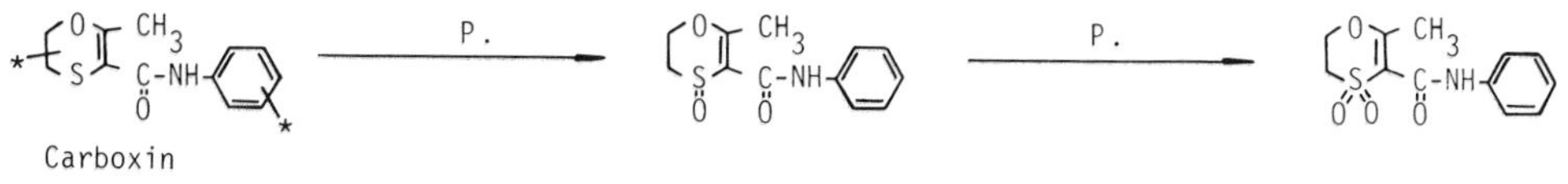

Plant Barley, Wheat[18]

TRIFORINE

In mammals and plants, the complete cleavage of one side chain of triforine has been observed, but stepwise degradation reactions take place during hydrolysis. The photodegradation product is *N*-2, 2-dichlorovinylformamide.

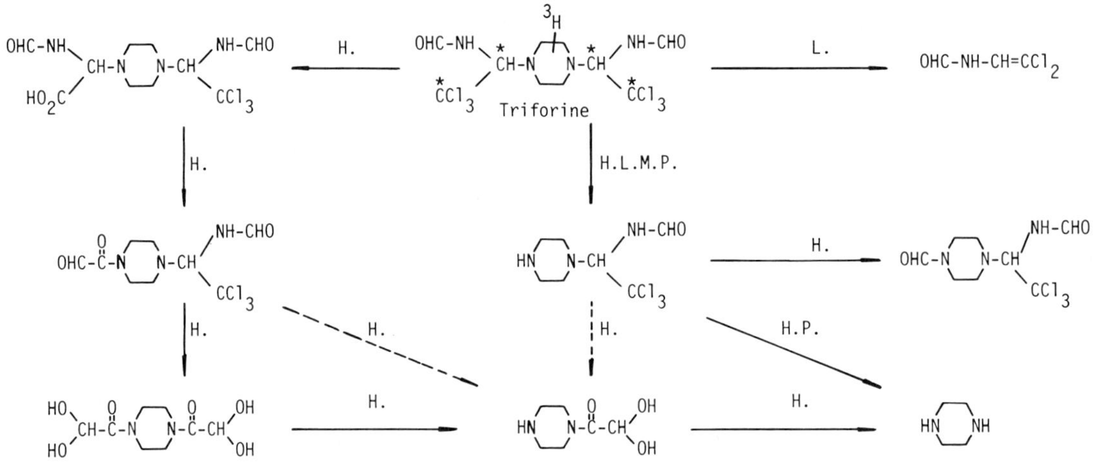

Plant	Barley: cv.Hebe[19]
Hydrolysis	pH 4.7, 6.8, 9.2
Light	Sunlight, UV, and high pressure lamp irradiation[20]
Mammal	Wistar rat: Male FW 49 Biberach strain[21]

PERFLUIDONE

The sulfonamide nitrogen of perfluidone is too hindered to form conjugates, but its tautomeric isomer with a more exposed reactive site might be expected to form an acid-labile conjugate. Acid-labile conjugates, water-soluble products, have been detected as conjugates in high concentration in the excised leaves. *N*-Hydroxy- and methylsulfonyl perfluidones have been detected in urine of rats and chickens, respectively.

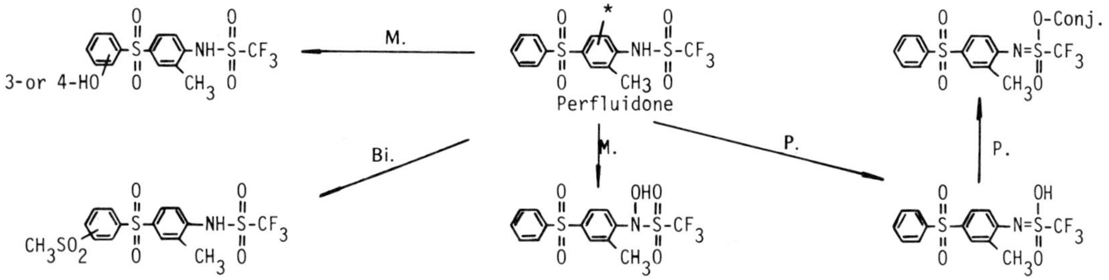

Mammal	Lactating Jersey cow[22]; Male rats[23]
Bird	Mature laying Leghorn hens[23]
Plant	Spanish peanut: *Arachis hypogaea* L.[24]

2

Amidines and Guanidines

GALECRON (CHLORODIMEFORM)

N,N-Disubstituted methyl groups are completely dealkylated by the usual decomposition reactions. Following normal metabolic patterns, galecron is decomposed to its deaminated formyl compound, *N*-formyl-4-chloro-*o*-toluidine, which is subsequently decomposed to the corresponding toluidine. The terminal metabolite is 5-chloroanthranilic acid.

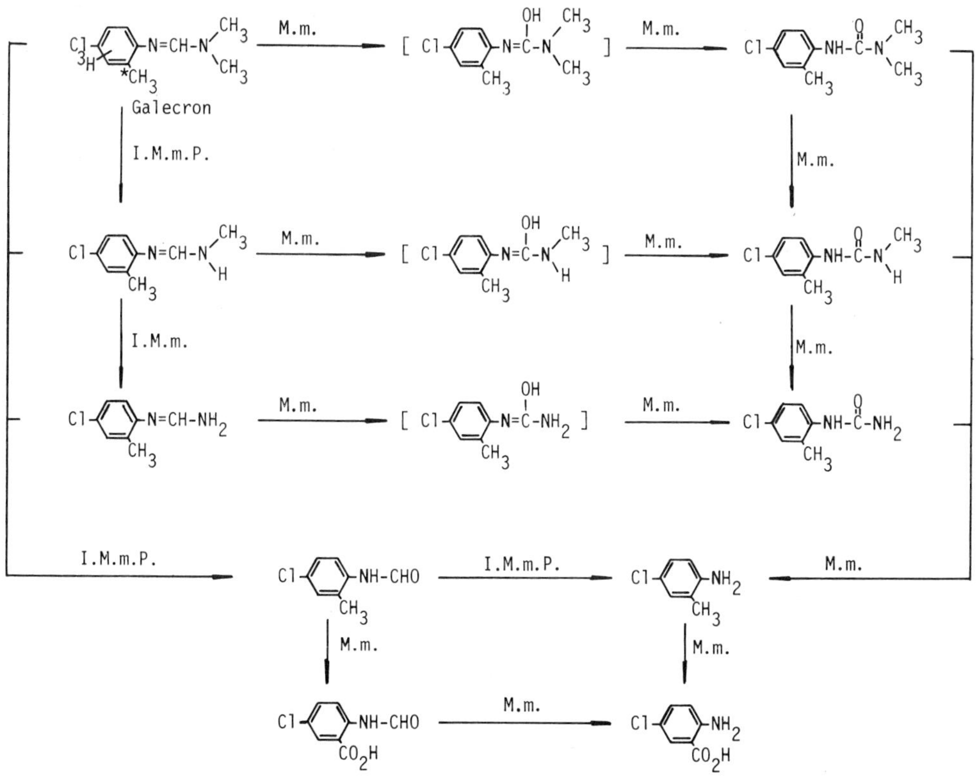

Plant	Dormant apple seedlings (Brooking variety)[25]
Insect	Two-spotted spider mites: *Tetranychus urticae* Koch[26]
Mammal	Swiss-Webster male mice[27]
Microsome	Rat liver microsomes[27]

ROBENZ (ROBENIDINE)

This rodenticide is easily split into *p*-chlorobenzoic acid and 1,3-diaminoguanidine by rodents themselves. Metabolism of aminoguanidine is now the problem. Toxicological evaluation of the metabolites is discussed in the references.

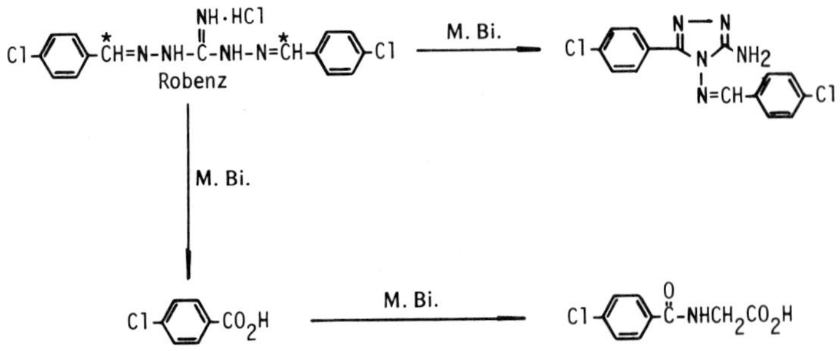

Mammal Male Charles River rats[28]

Bird Male Peterson-Cross chickens[29]

3

Anilines and Nitrobenzenes

p-CHLOROANILINE

p-Chloroaniline (not a pesticide) has no biological effect on agricultural infestations and domestic hazardous insects. However, this aniline is the important intermediate in manufacturing various pesticides and their environmental metabolites.

It is necessary to be able to detect *p*-chloroaniline metabolites when investigating the metabolism of pesticides that have the *p*-chloroaniline moiety in their structures. Generally anilines are transformed to N-acylated, ring-hydroxylated, and diazotized compounds. Triazene formation which may also occur with other anilines is observed in bacteria metabolism.

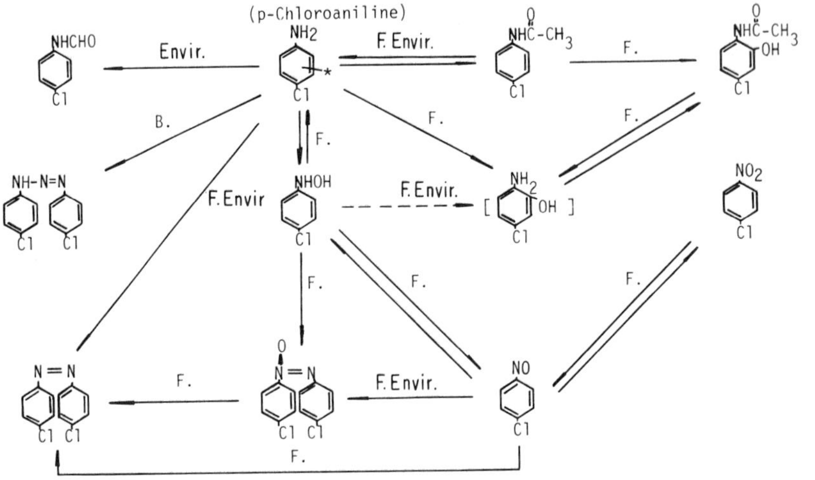

Fungi	*Fusarium oxysporum* Schlecht mass[30]
Bacteria	Gegnus *Paracoccus* sp.[31]
Environment	Alga; water[32]

PROPHAM

Various modes of metabolism are operative in different animals. Hydrolysis of the amide bond, hydroxylation of the phenyl ring and isopropyl group, and resulting conjugations are included in the metabolism of propham.

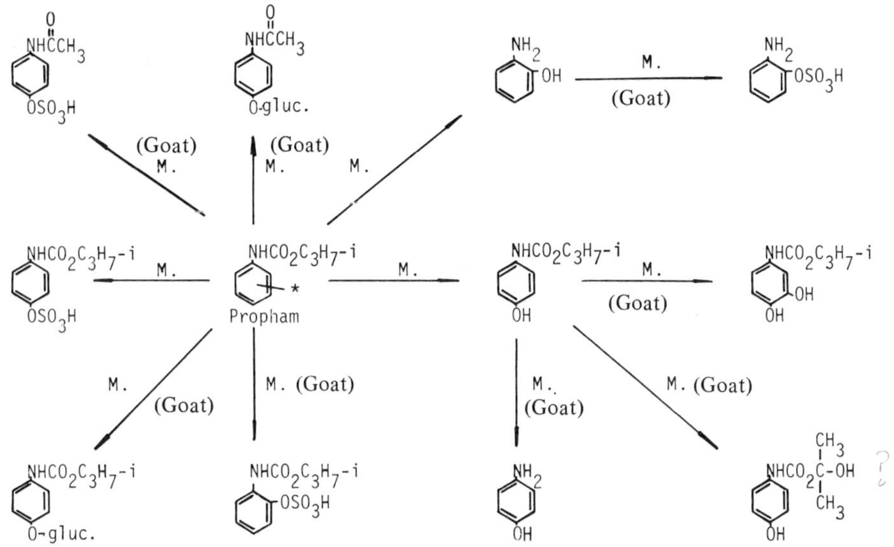

Mammal Male S-D rats; Lactating goat[33]

p-NITROANISOLE

Although *p*-nitroanisole is not a pesticide, its primitive metabolic pathway in microsomal systems is shown.

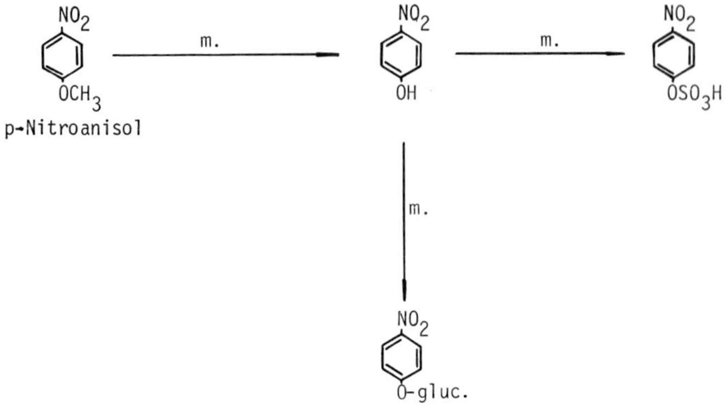

Microsome Male S-D rat liver microsomes[34]

CHLOROANISIDINE

Products identified in soil are consistent with a one-electron oxidation to a free radical, and N–N coupling gives a substituted hydrazobenzene which could be rapidly dehydrogenated to the azo compound. N-para coupling, followed by the loss of a methanol, yields a substituted *p*-quinone diimine that is rapidly hydrolyzed to ammonia and quinone monoimine, which is then reduced to the hydroxydiphenylamine.

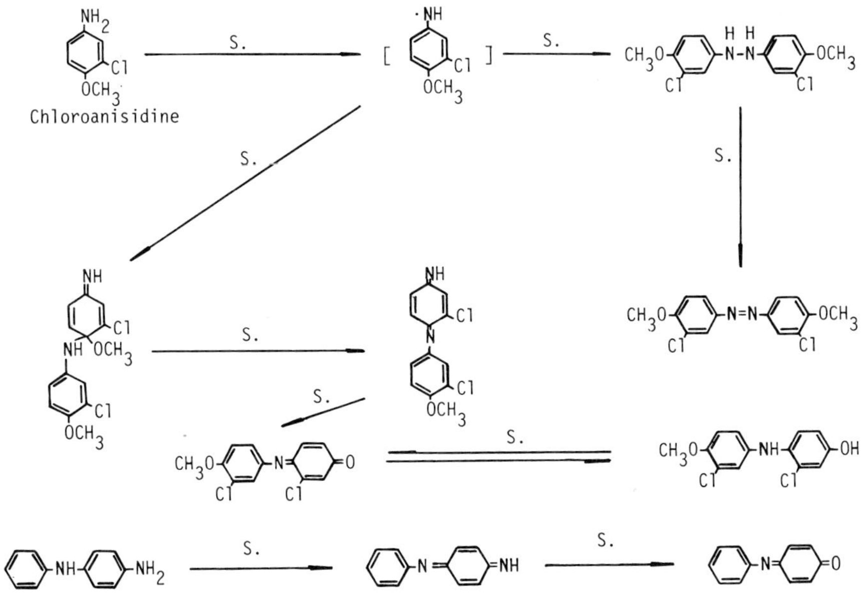

Soil Clay loam pH 8.1; Slurry pH 7.4[35]

DCNA (DICHLORONITROANILINE)

DCNA, which has amino and nitro groups at the 1 and 4 positions of the phenyl ring, is reduced to *p*-phenylenediamine; a more sterically hindered amino group has not been acetylated by bacteria.

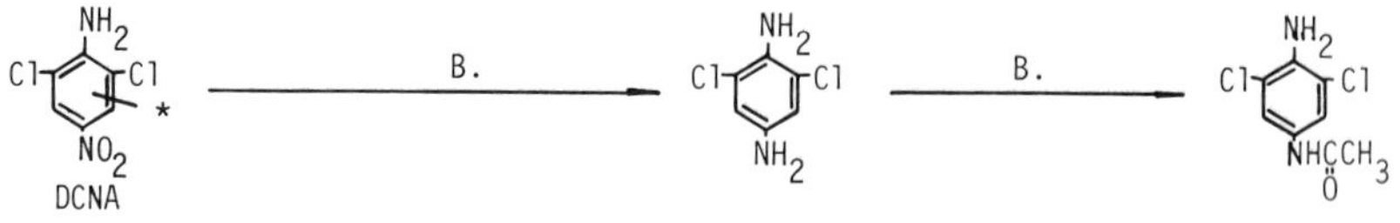

Bacteria *Escherichia coli* B; *Pseudomonas cepacia*[36]

DINITRAMINE

Dealkylation of the N,N-disubstituted alkylamino group occurrs in fish. During photodegradation, an *N*-alkyl group is cyclized with a less hindered nitrogen atom at the 4-position (from an amino group) to form benzimidazoles.

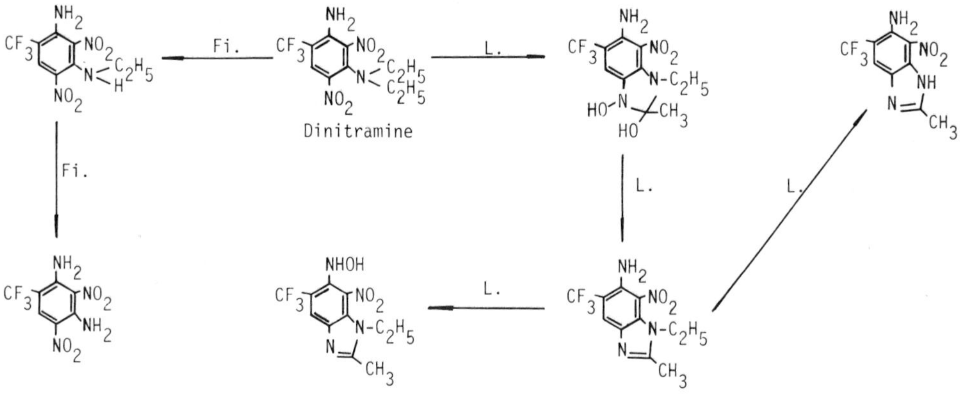

Light	Sunlight irradiation in natural water; UV light irradiation[37]
Fish	Carp: *Cyprinus carpio*[38]

DINOBUTON

The ester group of dinobuton is hydrolyzed to its dinitrophenol, which is further oxidized at the *sec*-butyl group to yield carboxylic acids. The two nitro groups are separately reduced to amines and in mammals, the *p*-amino group of the phenol is acetylated. In plants, the subsequent conjugation forms *O*-glucosides and/or *N*-glucosides.

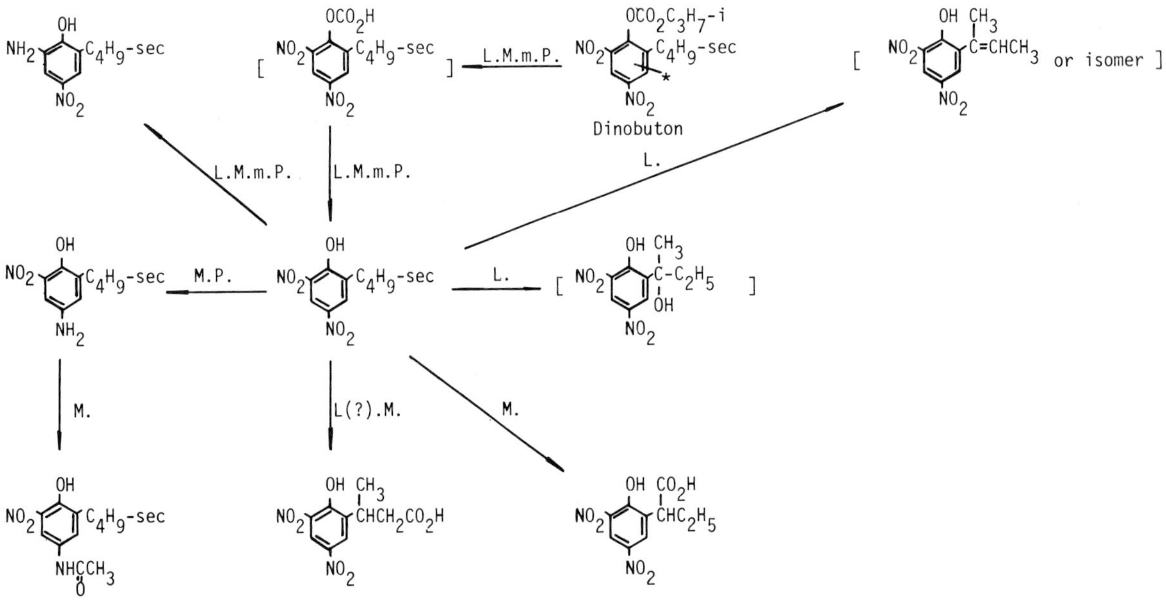

Light	Sunlight[39]
Mammal	Male albino rats and mice[39]
Microsomes	Male albino S-D rat liver microsomes; Housefly abdomens *Musca domestica* (R-Baygon strain)[39]
Plant	Bean: *Phaseolus vulgaris*[39]

BBD (*N-SEC*-BUTYL-4-*TERT*-BUTYL-2,6-DINITROANILINE)

In the course of photolysis of BBD, a radical at the tert-carbon atom is generated by hydrogen subtraction from the tert-carbon atom. This radical is then converted to the N-dealkylated compound, nitrosobenzene, and benzimidazoline in which one of the nitrogen atoms is hydroxylated. The hydroxylated product at the sec-carbon atom of an *N-sec*-butyl group is the metabolite in fungi.

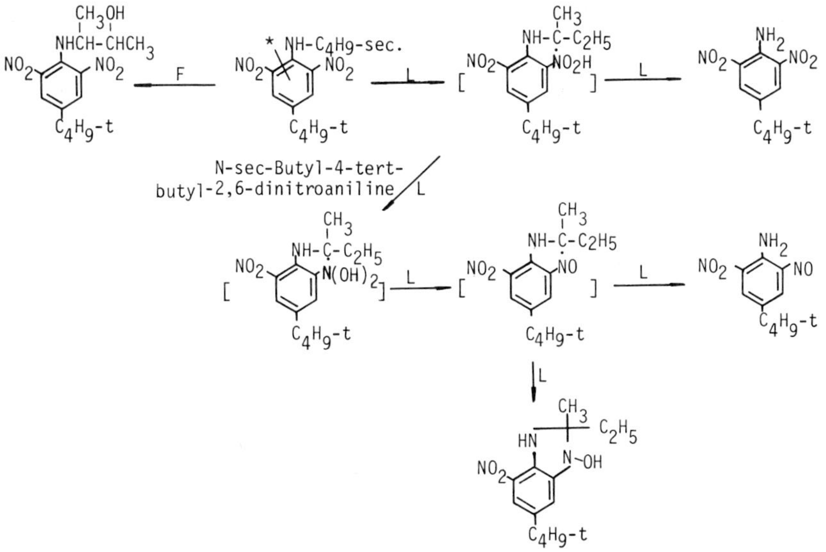

Light Natural sunlight; Sun lamp; Mercury vapor lamp[40]

Fungi A soil fungus: *Paecilomyces* sp.[41]

TRIFLURALIN

The predominant metabolic pathways appear to be hydroxylation of alkyl groups or N-dealkylation. To a lesser extent, a cyclized compound, benzimidazole, and the reduction product of a nitro group, an amine, are also included in the pathways. (Toxicological data of the metabolites are given by Nelson.[45])

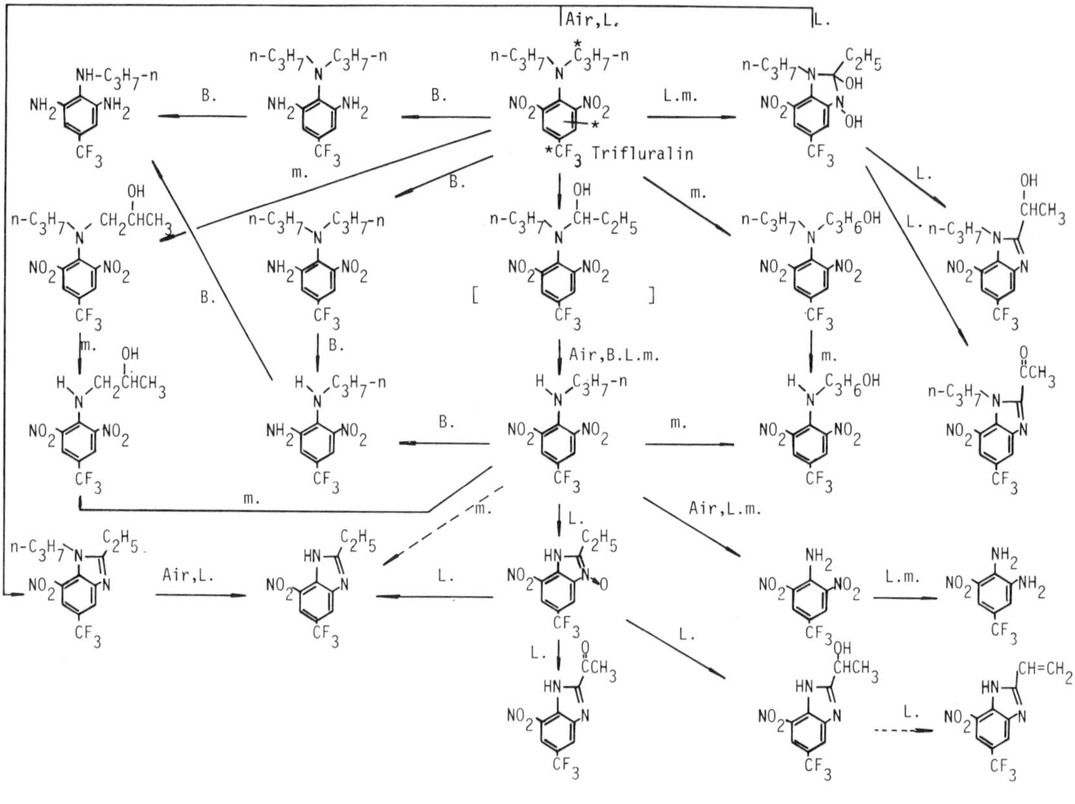

Bacteria

Rumen microorganisms: *Bacteroides amylophilus,* H-18; *B. ruminicola* subsp. *brevis,* GA-33; *B. succinogenes,* S-85; *Butyrivibrio fibrisolvens,* 49; *Eubacterium ruminanticum,* B_1C23; *E. limosum,* L-34; *Lachnospira multiparus,* D-32; *Peptostreptococcus elsdenii,* B-159; *Ruminicoccus flavefaciens,* C-94; *Streptococcus bovis,*

	FD-10; *Succinimonas amylolytica,* $B_2$4; *Succinivibrio dextrinosolvens,* 24[42]
Light	Vapor-phase photolysis[43]
Environment	Air[44]
Microsomes	Hepatic microsomes from normal and phenobarbital pretreated male S-D rats[45]

BASALIN

In light, basalin is degraded by dealkylation and then reduction to form nitrosobenzenes. Other cyclization reactions give unique products of benzimidazole analogs via radical formation.

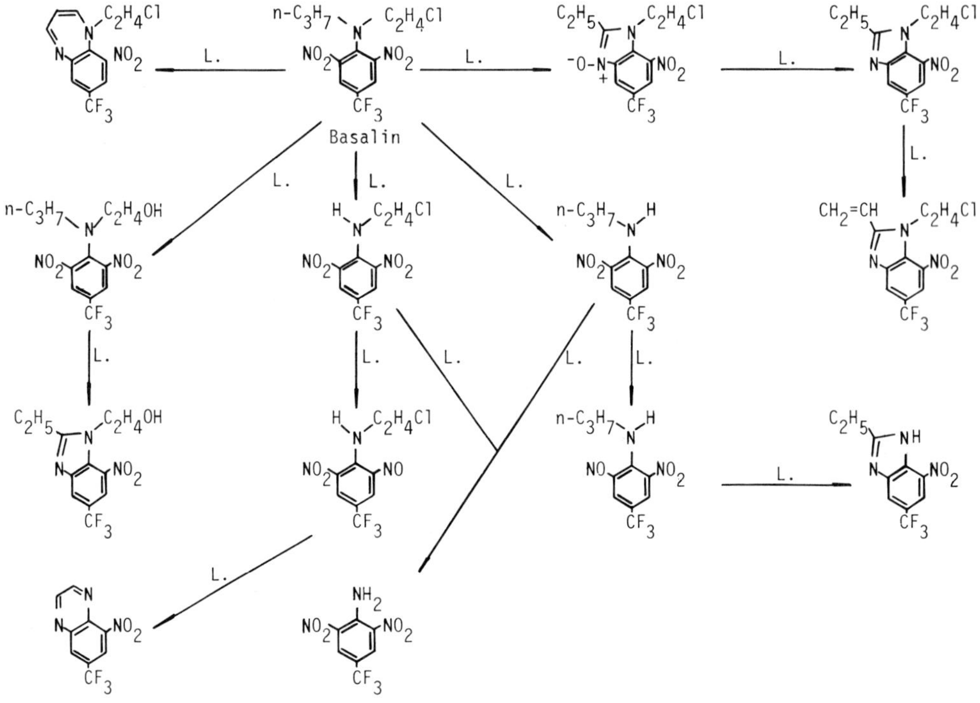

Light Photoreaction in solution, on soil (Montcalm sandy loam), on silica gel plate, and thin film[46]

PCNB (PENTACHLORONITROBENZENE)

The nitro group is replaced by OH or SCH_3 or reduced to NH_2, but the main metabolite is not known.

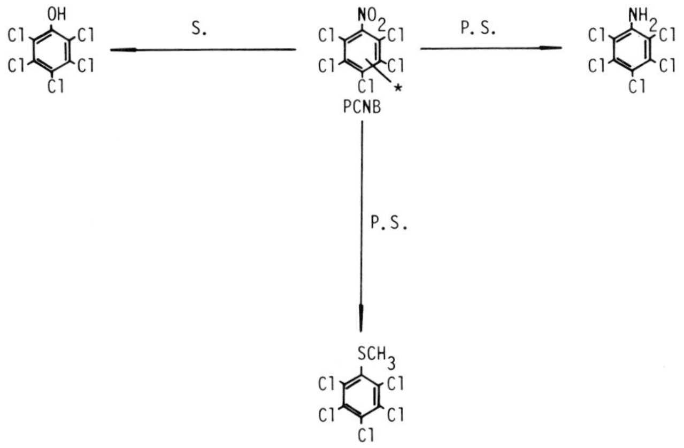

Plant	*Zea mays* L. golden cross bantan[47]
Soil	Anaerobic soil[48]

4

Biphenyl Ethers

NITROFEN, CNP, BIFENOX (MC-4379), OXYFLUOROFEN (RH-2915), CHLOROMETHOXYNYL (X-52), AND PREFORAN

Most of the *p*-nitro groups in one phenyl ring of these biphenyl ether herbicides (except preforan) are reduced to amines, followed by acetylation. Chlorine atoms substituted at the other phenyl rings are reduced or replaced by a hydroxyl group and, in some case, rearranged by hydroxylation according to NIH shifting (nitrofen). A cleavage of the ether bonds has been observed in nitrofen, CNP, preforan, and bifenox herbicides.

Nitrofen and CNP

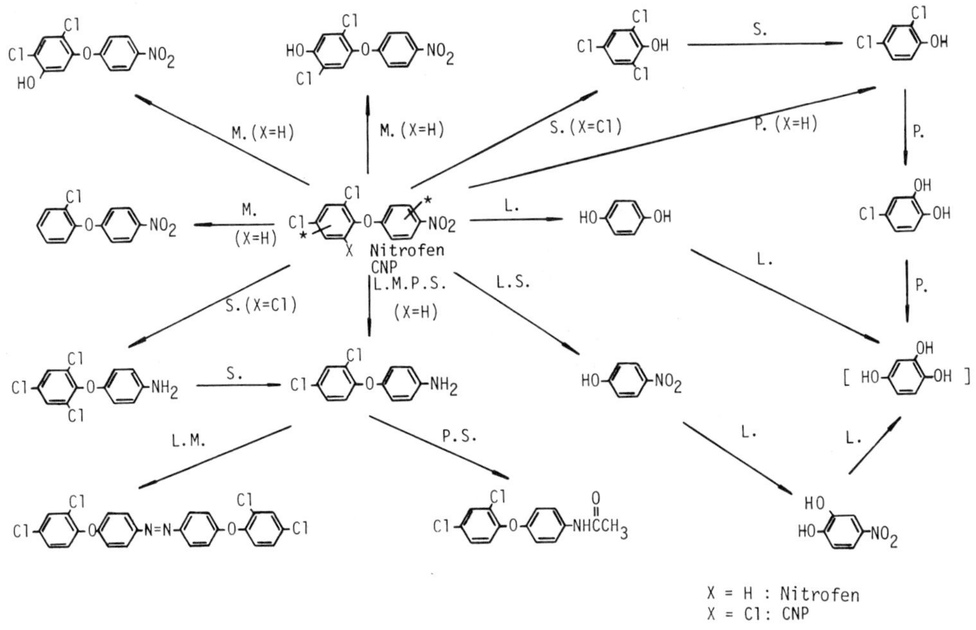

Light	Fluorescent lamp; Sunlight[49]
Plant	Rice and wheat grains[50]
Soil	[5]
Mammal	Delaine ewe[51]

Chloromethoxynyl (X-52)

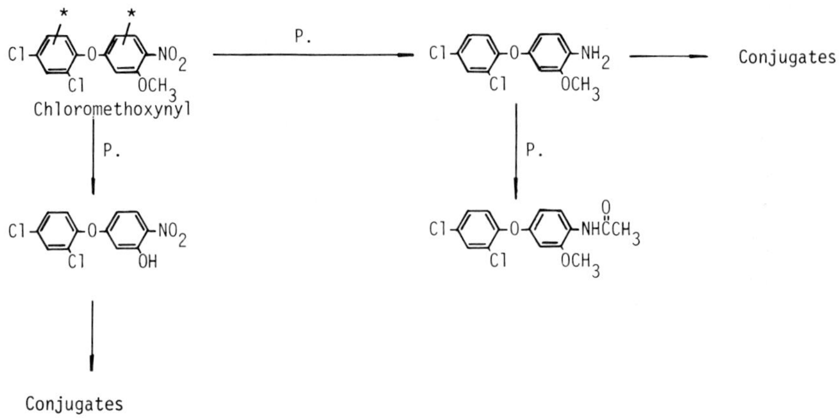

Plant Rice seeds: *Oryza sativa* L.; Barnyard millet: *Panicum crus-galli* L., var. *frumentaceum* Trin.[52]

Preforan

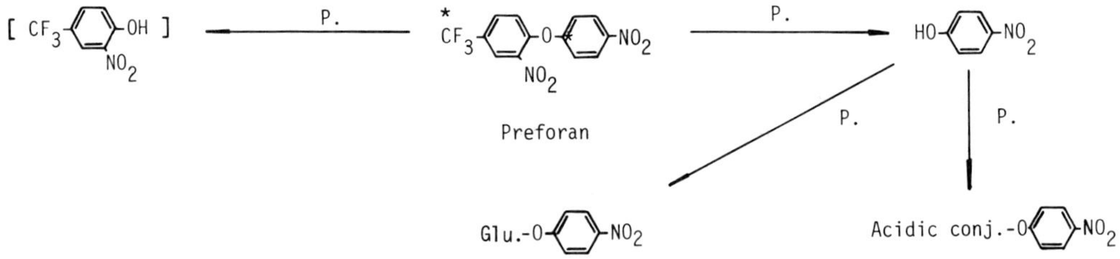

Plant XD cell line of *Nicotiana tabacum* L., var. Xanthi[53]

Bifenox (MC-4379)

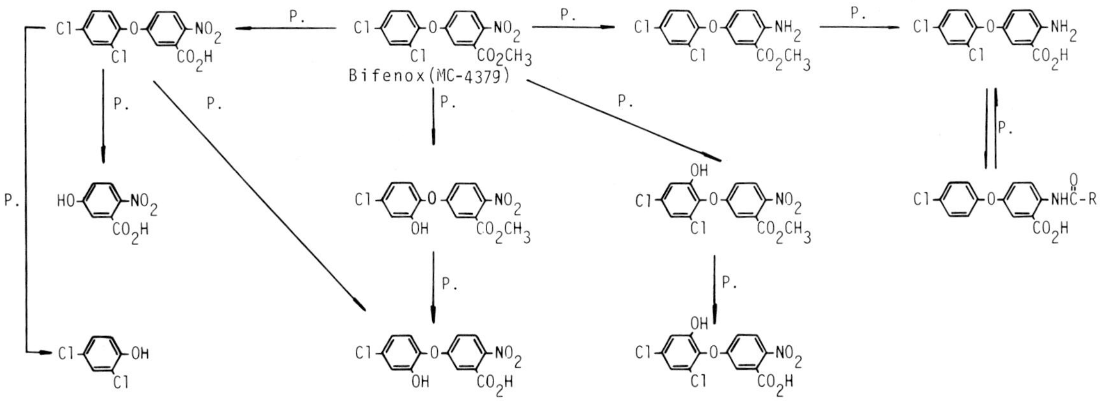

Plant 5

Oxyfluorofen (RH-2915)

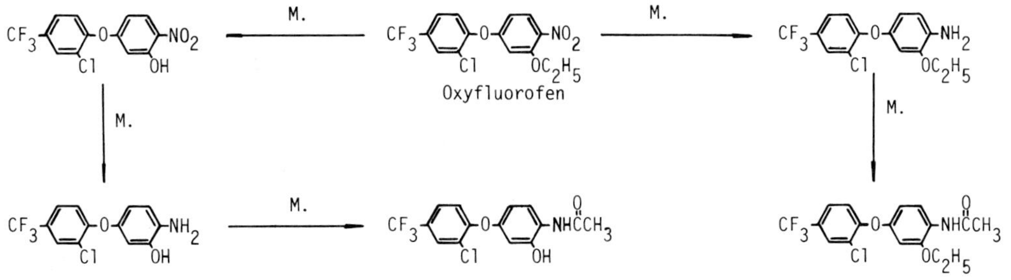

Mammal Female and male albino rats[54]

CREDAZINE (H-772)

p-Nitrobiphenyl ethers generally undergo cleavage of the ether bond by mixed function oxidase, but in the case of credazine, bond cleavage is promoted by nonenzymatic degradation after the methyl group is completely oxidized to carboxylic acid in mammal. Direct ether bond cleavage can be observed in plants.

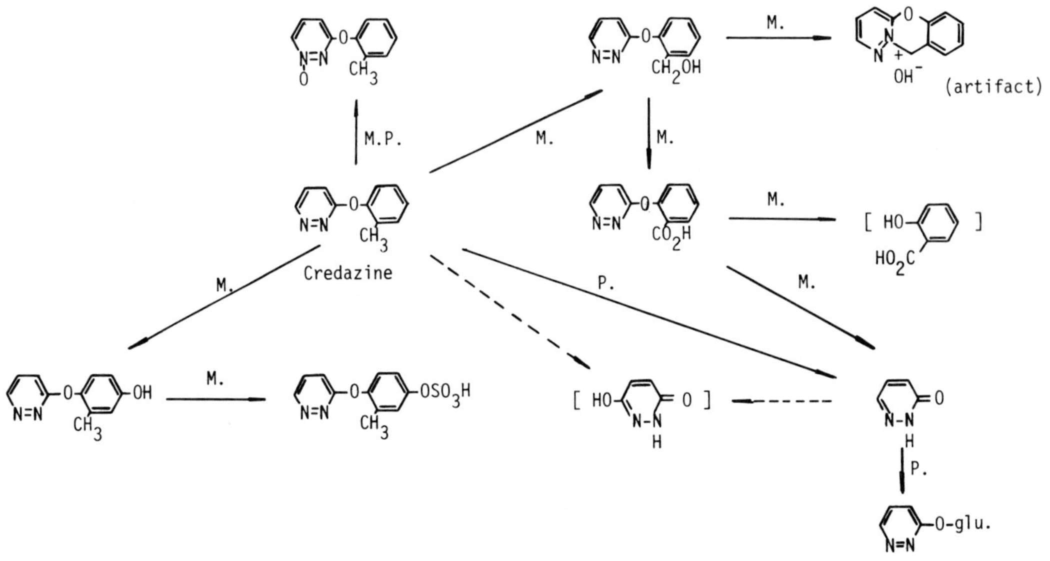

Plant Barley: *Hordeum vulgare* L. var. Akashinriki; Tomato: *Lycopersicon esculentum* Mill var. Fukuju No. 2[55]

Mammal Male Wistar rats[56]

5

DDT and Its Analogs

DIPHENYLMETHANE AND DTE (DIPHENYLTRICHLOROETHANE)

In comparison with the chlorine-substituted diphenylmethanes such as DDT (p. 39), diphenylmethane and DTE are rather easily oxidized. Cleavage of the phenyl ring gives acetic acids via benzylcatechols especially by *Hydrogenomonas* sp. The difference in degradability is dependent on species specificity and strongly on the chemical properties of chlorine-substituted or unsubstituted diphenylmethanes.

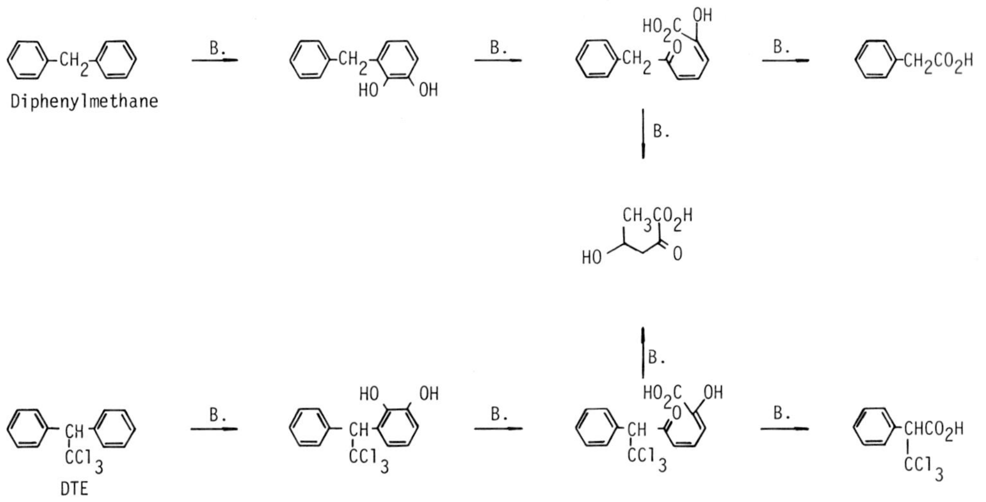

Bacteria *Hydrogenomonas* sp. (co-metabolism)[57]
Pseudomonas putida (co-metabolism)[58]

DDT [1,1,1-TRICHLORO-2,2-BIS (*p*-CHLOROPHENYL)ETHANE]

DDT metabolic pathways are well known and occur in various precisely investigated metabolic systems.[59a]

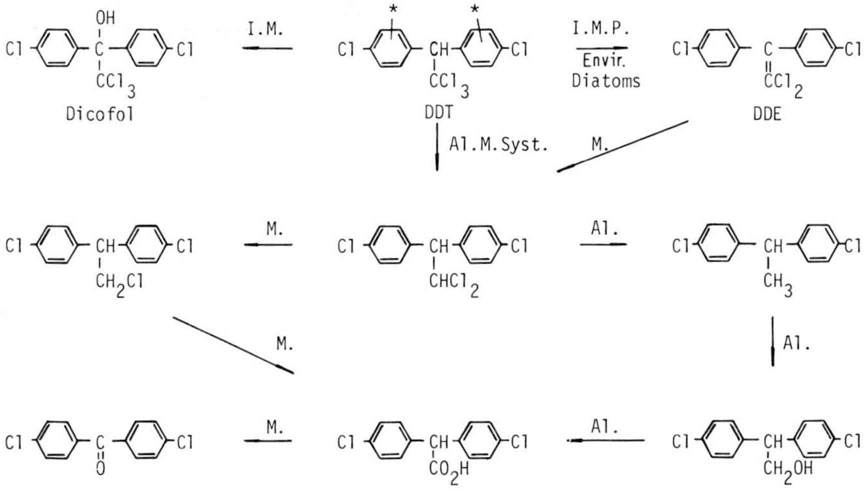

Mammal	Female Swiss mice[59b] Male Swiss mice[60]
Insect	Female SP (DDT resistant) housefly; Fifth instar larvae of saltmarsh caterpillar, *Estigmene acrea*[59b]
Environment	A model ecosystem = Plant: *Sorghum* halpense; Algae: *Oedogonium cardiacum;* Snails: *Daphnia magna, Physa;* Larvae of the salt marsh caterpillar: *Estigmene acrea;* Mosquito larvae: *Culex quinquefasciatus;* Fish: *Gambusia affinis*[59b]
Diatoms	*Nitzschia* sp.[61]
Algae	Oceanic microorganisms *Dunaliella* sp.; *Agmenellum quadraplicatum* (strain PR-6); Field collected algae containing water samples[62]
Plant	Marine algae *Skeletonema costatum, Cyclotella nana* (Bacillariophyta), *Amphidinium eartei* (Pyrrophyta), *Olisthodiscus* luteus (Xanthophyta), *Tetraselmis chuii* (Euglenophyta)[63]

DDE [1,1′-(DICHLOROETHENYLIDENE) BIS[4-CHLOROBENZENE]]

DDE, a dehydrochlorinated product of DDT, is an end product in some metabolic systems. Although the C=C double bond is not very stable in metabolic degradation reactions, in the mammalian condition described in ref. 64, one phenyl ring is selectively epoxidized and then hydroxylated, and all the metabolites including an NIH-shifted metabolite, which retains a C=C double bond in its structure, can be detected, but conversion to these metabolites is low.

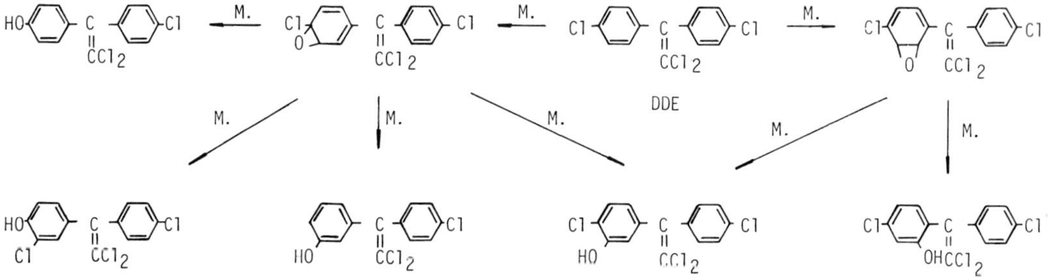

Mammal Rat[64]

o,p'-DDT [1,1,1-TRICHLOROPHENYL-2-(*o*-CHLOROPHENYL)-2-(*p*-CHLOROPHENYL)ETHANE]

A phenyl ring with chlorine substituted at the ortho position is selectively hydroxylated, but the chlorine atom itself is not eliminated by any oxidation reaction. A conjugated metabolite has been identified as a serine conjugate of *o,p'*-dichlorodiphenylacetic acid.

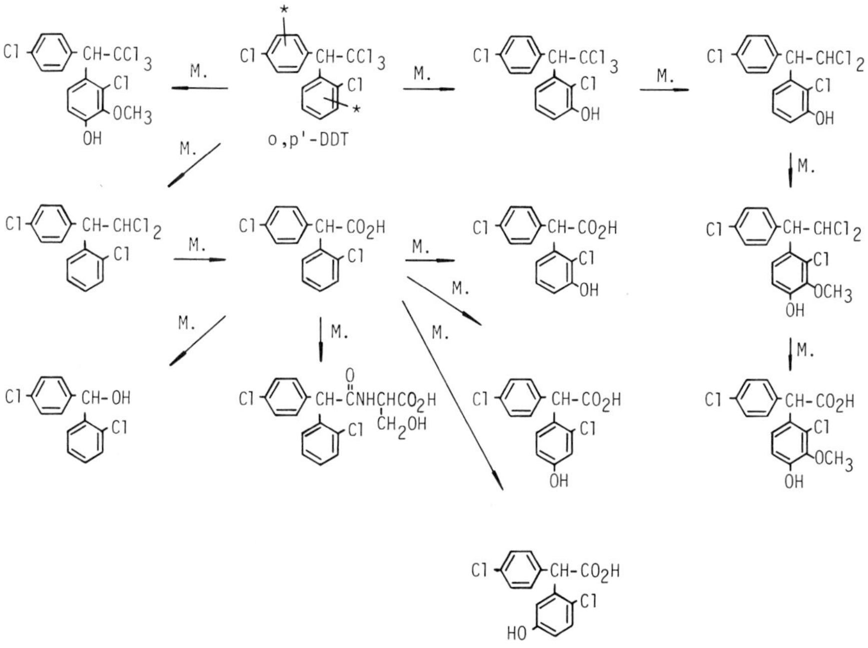

Mammal Rats[65]

DDA [BIS(*p*-CHLOROPHENYL) ACETIC ACID] AND DDM [BIS(*p*-CHLOROPHENYL)METHANE]

DDA is converted to DDM, which is oxidized to cleave the phenyl ring yielding 3-chloro-4-hydroxy-2-ketopentanoic acid and *p*-chlorophenylacetic acid selectively by *Pseudomonas putida.* The latter is further degraded to glycol aldehyde by *Hydrogenomonas* sp. This bacterium also converts DDM to the terminal metabolite, a benzophenone derivative.

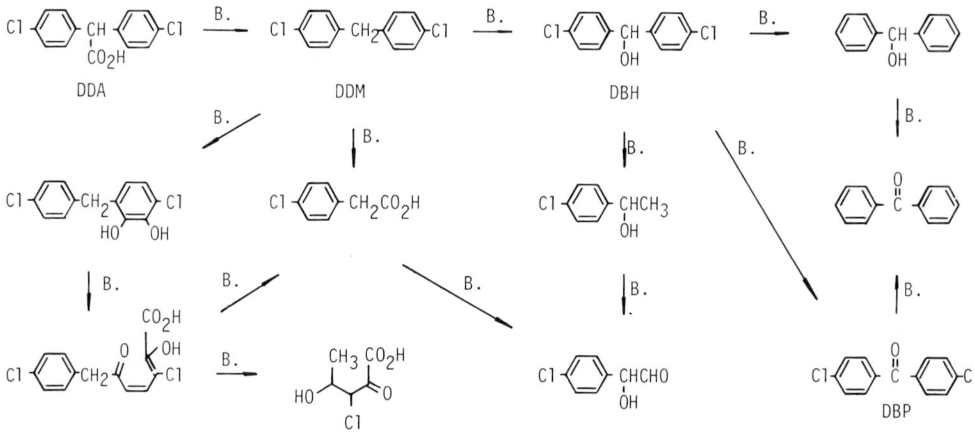

Bacteria *Hydrogenomonas* sp. (co-metabolism)[57]
Pseudomonas putida (co-metabolism)[58]

CHLOROPROPYLATE

Chloropropylate is hydrolyzed to its corresponding carboxylic acid, which is further conjugated in mammals.

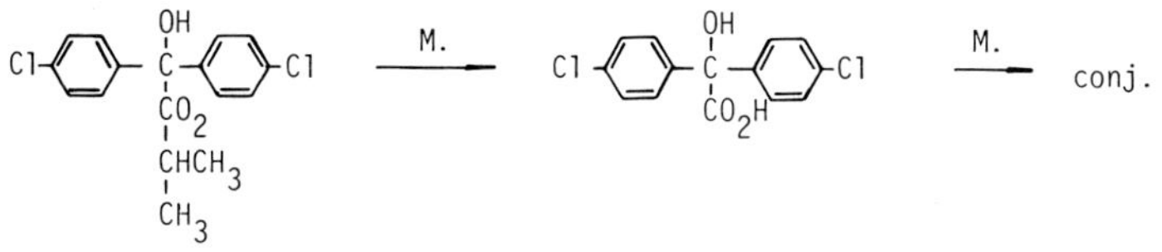

Chloropropylate

Mammal Holstein cow[66]

CHLOROMETHYLCHLOR

Oxidation of a methyl group to carboxylic acid and dehydrochlorination followed by oxidation to benzophenones comprise the metabolic pathways of chloromethylchlor. Dechlorination and phenyl ring oxidation are not observed because of the predominant oxidation of the methyl group.

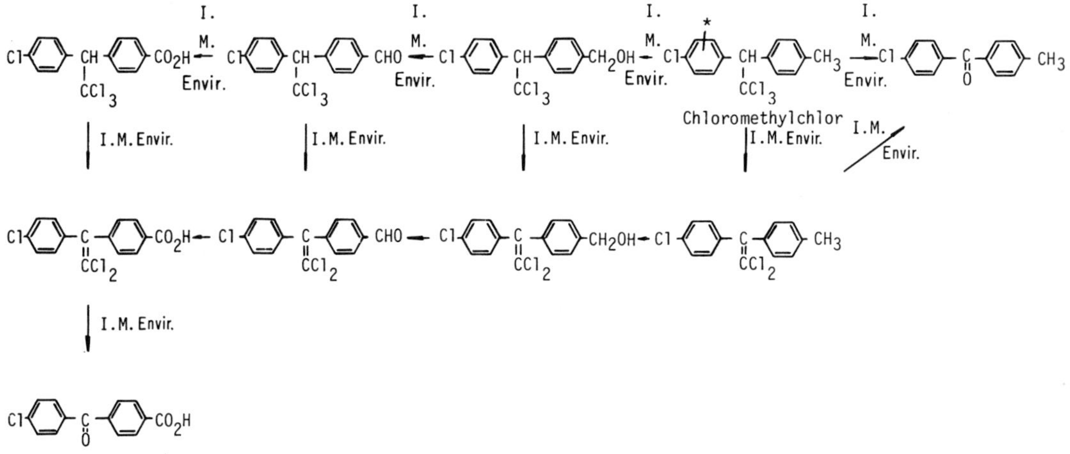

Insect	Saltmarsh caterpillar[67]
Mammal	Male Swiss white mice[67]
Environment	A model ecosystem[67]

METHYLCHLOR

The two methyl groups of methylchlor are stepwise oxidized to *p,p'*-dicarboxyphenyltrichloroethane; this is followed by dehydrochlorination to form a metabolite of the ethylene analog. The other pathway gives a directly dehydrochlorinated ethylene analog, a DDE-type compound.

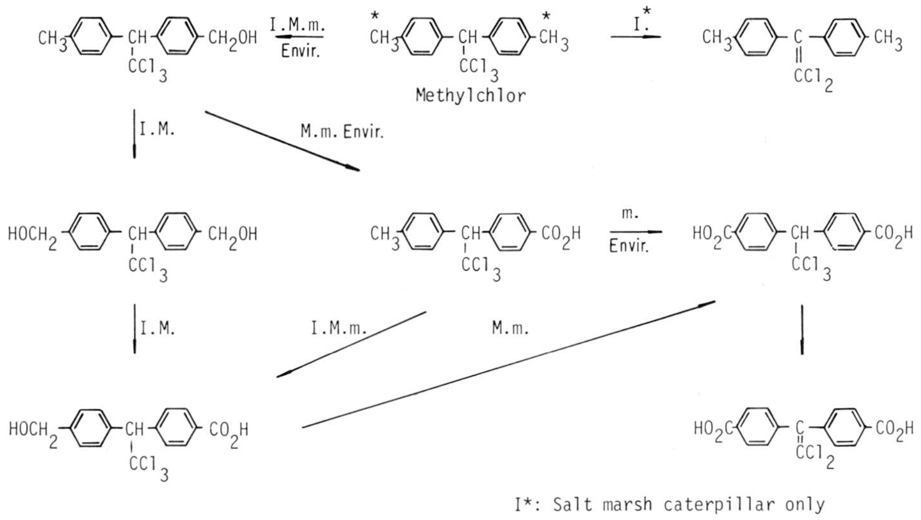

Mammal	Male Swiss mice[60]
Microsome	Mouse liver homogenate[60]
Insect	Female SP (DDT resistant) housefly; Fifth instar larvae of the salt marsh caterpillar: *Estigmene acrea*[60]
Environment	A model ecosystem = Plant: *Sorghum halpense;* Algae: *Oedogonium cardiacum;* Snail: *Daphnia magna* and *Physa;* Larvae of the salt marsh caterpillar: *Estigmene acrea;* mosquito larvae: *Culex quinquefasciatus;* Fish: *Gambusia affinis*[60]

METHIOCHLOR

Both sulfur atoms of methiochlor undergo oxidation, which leads to bissulfoxide and bissulfone. The ethylene derivative with no oxidized sulfur atoms results from dehydrochlorination of methoxychlor.

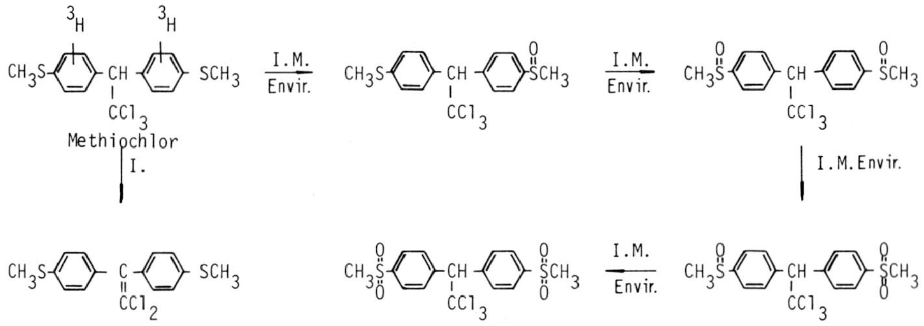

Mammal	Female Swiss mice[59]
Insect	Female SP (DDT resistant) housefly; Fifth instar larvae of salt marsh caterpillar: *Estigmene acrea*[59]
Environment	A model ecosystem = Plant: *Sorghum halpense;* Algae: *Oedogonium cardiacum;* Snails: *Daphnia magna, Physa;* Larvae of the salt marsh caterpillar: *Estigmene acrea;* Mosquito larvae: *Culex quinquefasciatus;* Fish: *Gambusia affinis*[59]

ETHOXYANILINE

Compared to ethoxychlor (p. 50), ethoxyaniline is more biodegradable because of the existence of a C–N bond in its structure. Degradation compounds which are similar to those of TMMA (p. 52) are produced in ethoxyaniline metabolism.

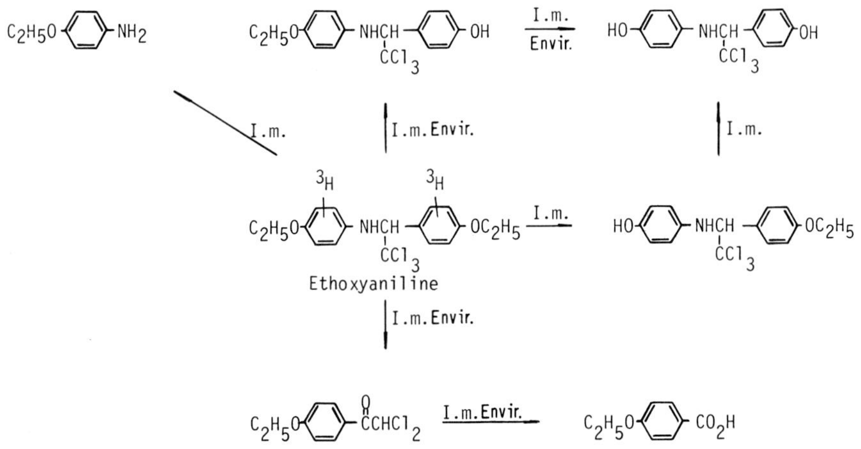

Insect	Adult female housefly; Larval salt marsh caterpillar: *Estigmene acrea* Drury[68]
Microsome	Mouse liver homogenate[68]
Environment	A model ecosystem = Algae: *Oedogonium;* Snail: *Physa;* Mosquito: *Culex;* Fish: *Gambusia*[68]

METHYLETHOXYCHLOR

Dehydrochlorination to form a DDE-type compound, hydrolysis of the ethyl ether bond to form a phenolic compound, followed by dehydrochlorination to produce another DDE-type compound, and oxidation of methyl group to carboxylic acid are observed in the metabolism of methylethoxychlor.

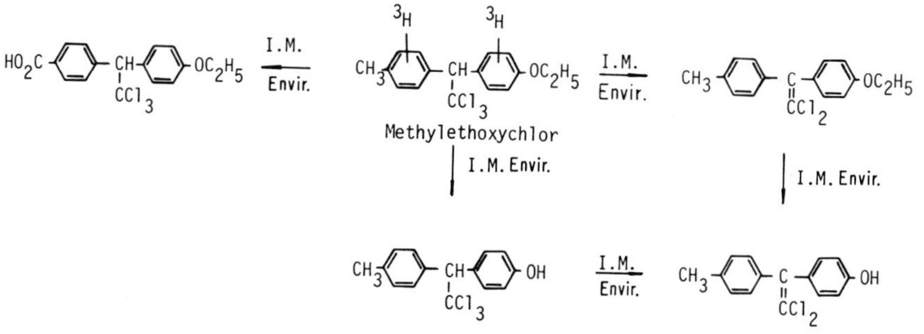

Insect	Salt marsh caterpillar[67]
Mammal	Male Swiss white mice[67]
Environment	A model ecosystem[67]

METHOXYCHLOR

Besides direct dehydrochlorination of methoxychlor, complete hydrolysis of the methyl ether bonds gives *p,p'*-dihydroxyphenyltrichlorethane, which is further degraded to produce *p,p'*-dihydroxybenzophenone in mammals.

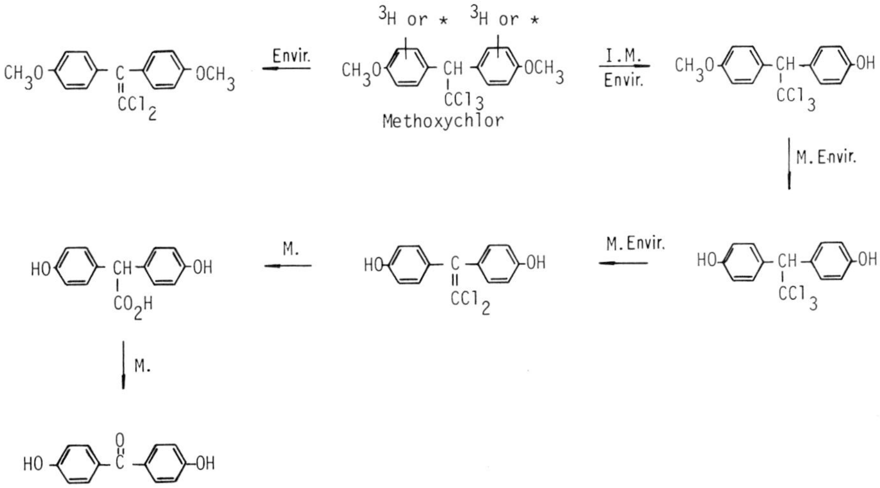

Mammal	Female Swiss mice[59]
Insect	Female SP (DDT resistant) housefly; Fifth instar larvae of salt marsh caterpillar: *Estigmene acrea*[59]
Environment	A model ecosystem = Plant: *Sorghum halpense;* Algae: *Oedogonium cardiacum;* Snails: *Daphnia magna, Physa;* Larvae of the salt marsh caterpillar: *Estigmene acrea;* Mosquito larvae: *Culex quinguefasciatus;* Fish: *Gambusia affinis*[59]

ETHOXYCHLOR

The degradation process for ethoxychlor is almost the same as that of methoxychlor with the additional hydroxylation of the ethyl group.

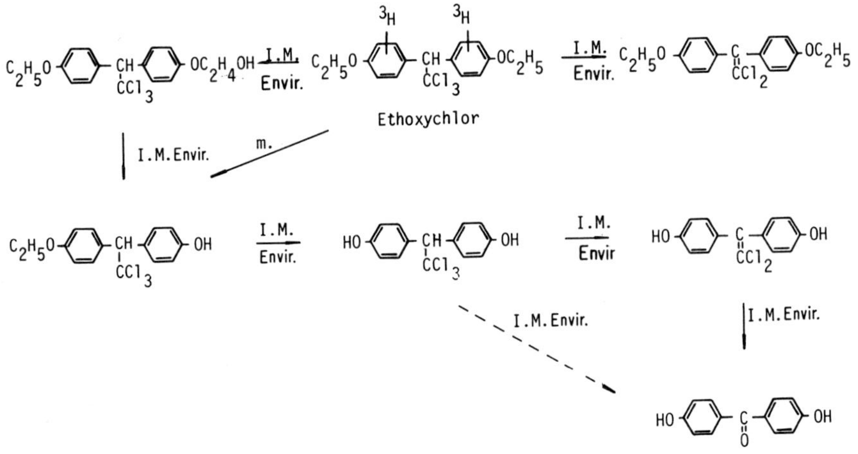

Mammal	Male Swiss mice[60]
Microsom	Mouse liver homogenate[60]
Insect	Female SP(DDT resistant) housefly; Fifth instar larvae of the salt marsh caterpillar: *Estigmene acrea*[60]
Environment	A model ecosystem=Plant: *Sorghum halpense;* Algae: *Oedogonium cardiacum;* Snails: *Daphnia magna* and *Physa;* Larvae of the salt marsh caterpillar: *Estigmene acrea;* Mosquito larvae: *Culex quinquefasciatus;* Fish: *Gambusia affinis*[60]

METHOXYMETHIOCHLOR

Oxidation of the sulfur atom of methoxymethiochlor (although this reaction detoxifies this insecticide) and/or dehydrochlorination gives the terminal metabolite, which still possesses an intact methoxy group.

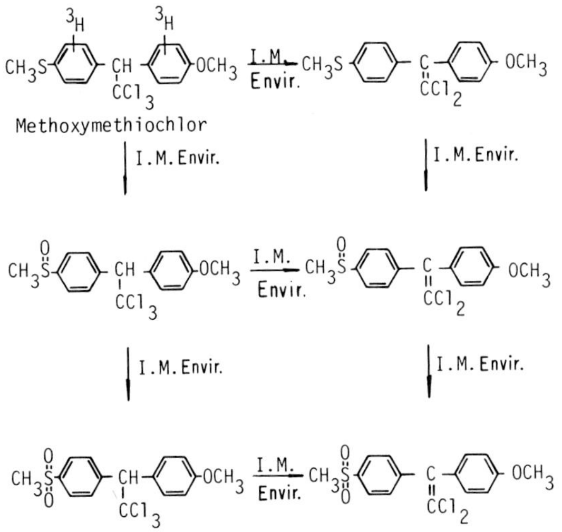

Insect	Saltmarsh caterpillar[67]
Mammal	Male Swiss white mice[67]
Environment	A model ecosystem[68]

TMMA [N-(α-TRICHLOROMETHYL-p-METHOXYBENZYL)-p-METHOXYANILINE]

TMMA is considerably more reactive than DDT under photolysis. TMMA is photodecomposed to give a rearranged product of mandelic acid amide as the major degradation product, and other fragments such as *p*-anisidine, acetophenone, benzaldehyde, and benzoic acid derivatives, and α,α,α′,α′-tetrachloroacetone.

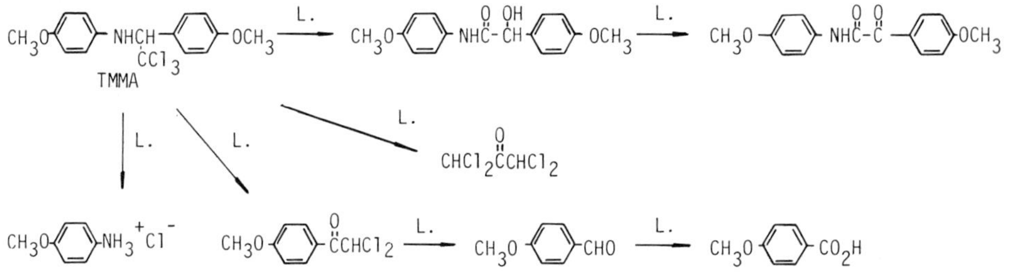

Light Photolysis[69]

6

Dithio- and Thiolcarbamates

ETHYLENE BIS-DITHIOCARBAMIC ACID

There are many degradative metabolites of ethylene bis-dithiocarbamic acid, including ethylene thiourea (ETU). ETU is further decomposed by hydrolysis or in plants and soil.

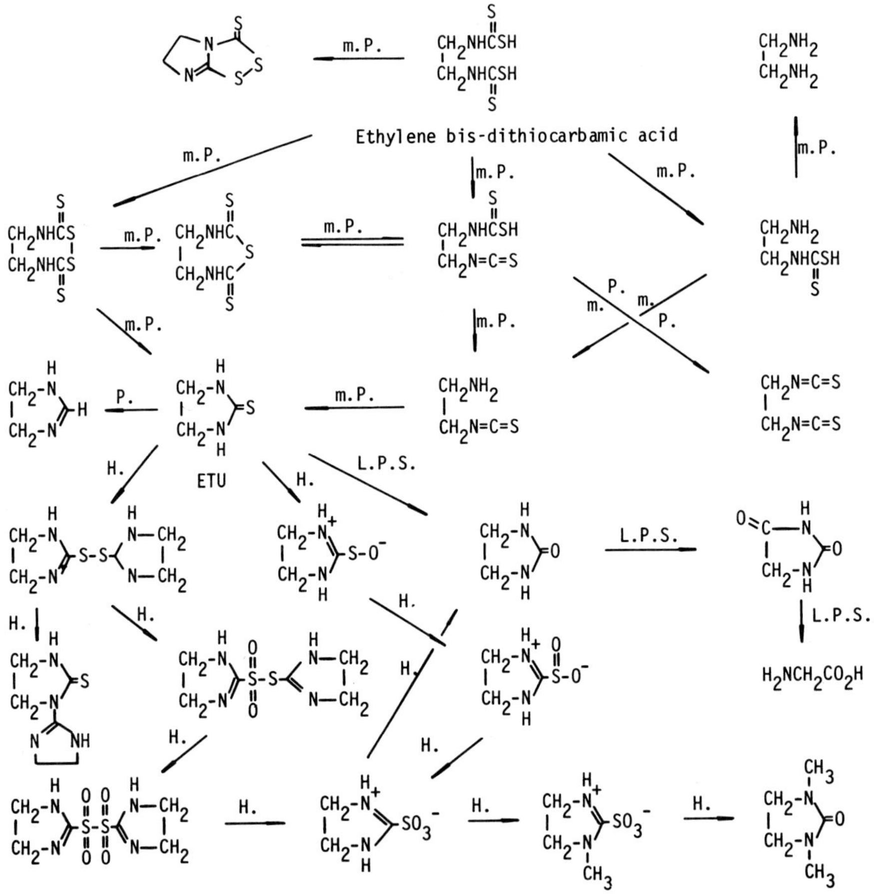

Microsome	*In vitro*[70]
Light	UV light[71]
Plant	Snapbean: *Phaseolus vulgaris,* var. Harvester; Tomato: *Lycopersicon esculentum,* var. Manalucie[71]
Soil	Soils[71]
Hydrolysis	NaOCl, H_2O_2 (1% $NaHCO_3$, 0.1 *N* HCl), I_2[72]

EPTC (S-ETHYL DIPROPYLTHIOCARBAMATE) AND BUTYLATE (*S*-ETHYL DIISOBUTYLTHIOCARBAMATE)

Although the yield of mercapturic acids is greater with EPTC than with butylate, the latter forms a greater number of metabolites. Cysteine conjugates are detected only from EPTC, whereas *N*-dealkylmercapturic acid is a major metabolite of butylate, but not of EPTC. These differences may result from a susceptibility for carbon hydroxylation in compounds with branched alkyl substituents on the nitrogen.

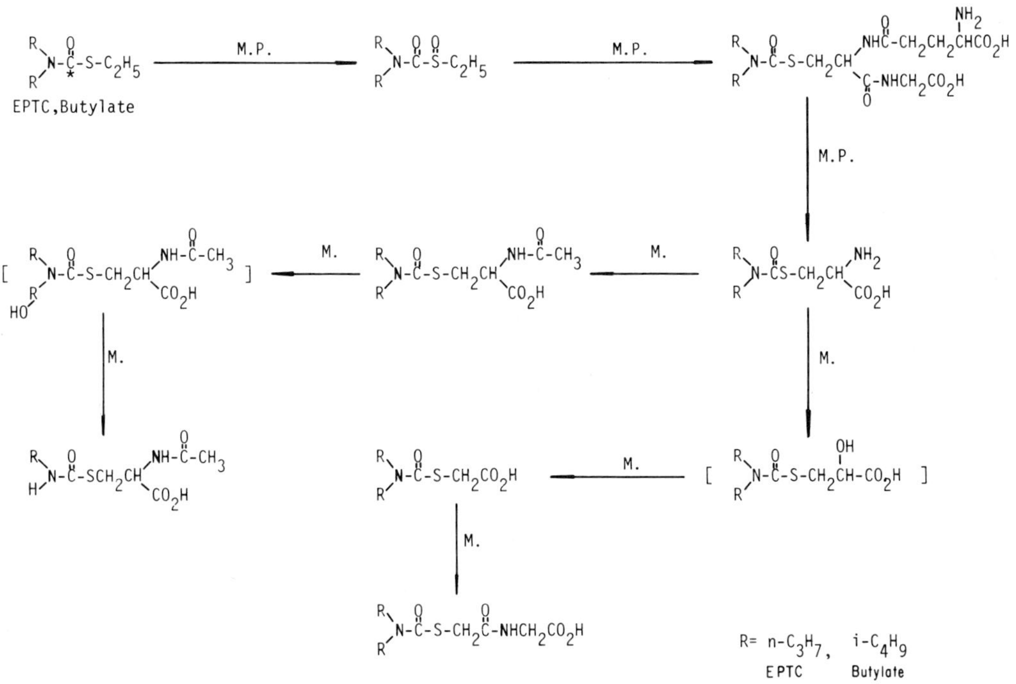

Plant	Corn: DeKalb, XL-66-H[73]
Mammal	Male albino S-D rats[73]

BENTHIOCARB

In addition to N-dealkylation and hydrolysis of the carbamoyl moiety which results in the formation of various benzyl alcohols, benzaldehydes, and benzoic acids, oxidation of the sulfur atom of the mother compound and coupling of radicals (*p*-chlorobenzylamine formation) has been observed.

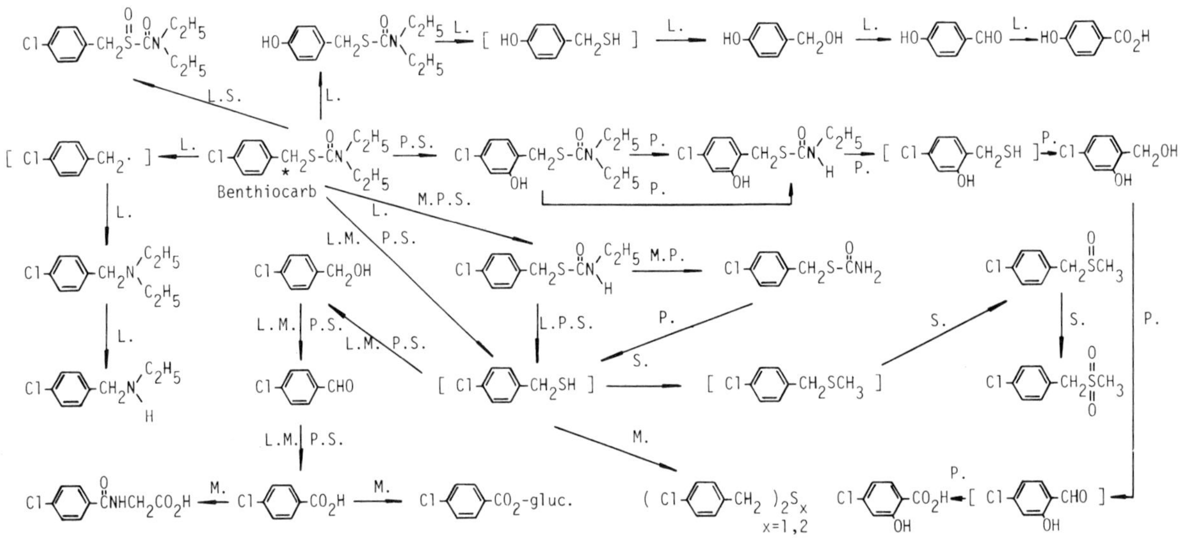

Mammal	Male mice, dd-strain[74]
Soil	[5,75]
Light	Low pressure mercury lamp[76]
Plant	[5]

MOLINATE (ORDRAM)

Sulfoxidation is presumed to occur prior to liberation of hydroxyimine analogs. A minor metabolite of a cyclic carbamate may be formed by internal rearrangement of 3-hydroxymolinate or its sulfoxide.

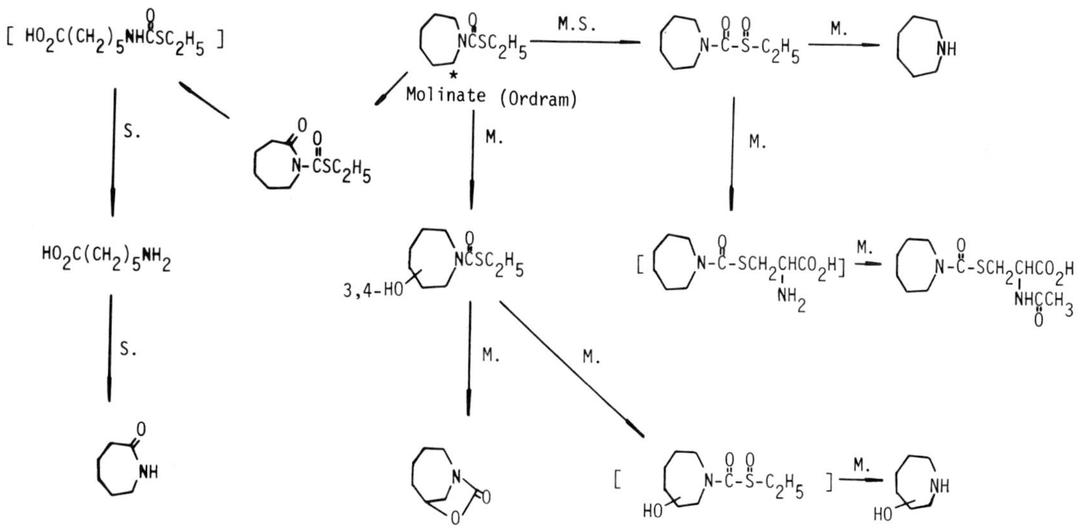

Soil Rice field (Sacramento clay soil)[77]

Mammal Female and male Simonsen albino rats[78]

7

Five- and Six-Membered Heterocyclic Compounds

PYRROLNITRIN

The pyrrole ring seems to be particularly susceptible to oxidation, which may account for the rapid degradation of pyrrolnitrin *in vivo*. The formation of a substituted maleimide by microsomal oxidation is most interesting in light of the known reactivity of maleimides with proteins and SH groups. Pyrrolnitrin is not a pesticide.

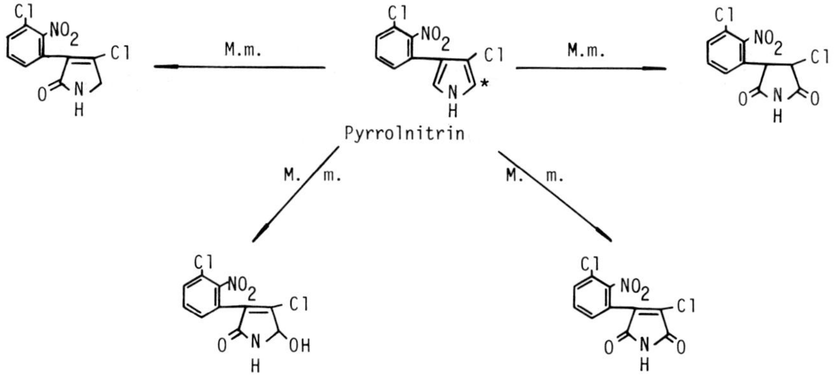

Mammal	Rat[79]
Microsome	Rat liver microsomes[79]

TACHIGAREN (F-319, HYMEXAZOL)

In plants F-319 is isomerized to its tautomer, isoxazolidone, which is also conjugated with glucoside.

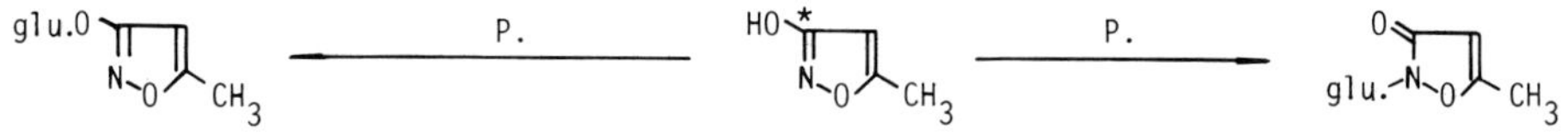

Tachigaren
(F-319)

Plant Cucumber: *Cucumis sativus* L. 'Midorifushinari'; Tomato: *Lycopersicon esculentum* Mill, 'Shinfukuju'; Rice: *Oryza sativa* 'Kinmaze' and 'Shigaasahi No. 27'[80]

IPRONIDAZOLE

The tertiary carbon atom of the substituted isopropyl group of the imidazole ring is oxidized in only microsomal systems.

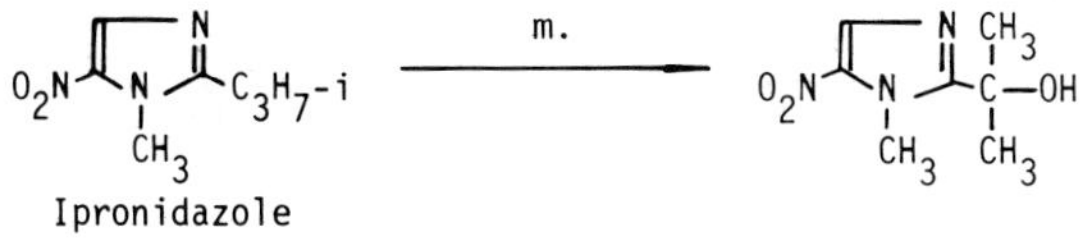

Microsome Turkey tissue[81]

ISOTHIAZOLINONES

Oxidation of sulfur atoms in the isothiazolinone ring followed by rapid ring opening leads to the degradative compounds, in the absence of a complexing agent like $CaCl_2$.

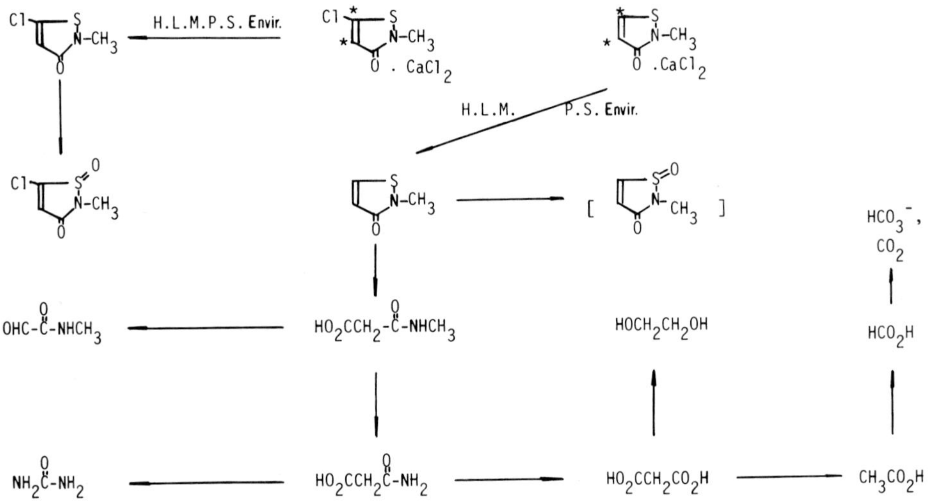

Hydrolysis	A basic hydrolysis system[82]
Light	A photolysis system[82]
Mammal	Rat urine and feces[82]
Plant	Aquatic plants[82]
Soil	An activated sludge system[82]
Environment	A river water system; An acetone/water system[82]

DDOD [3-(3,5-DICHLOROPHENYL)-5,5-DIMETHYL-2,4-OXAZOLIDINEDIONE, SILEX]

Ring opening at the $\overset{\mathrm{O}}{\overset{\|}{\mathrm{C}}}$-O bond of the oxazolidine ring and oxidation of the 4-position of phenyl ring are observed in the initial step of metabolism in plants, soil, and mammals. The N-substituted carboxyl group is easily eliminated by generation of CO_2.

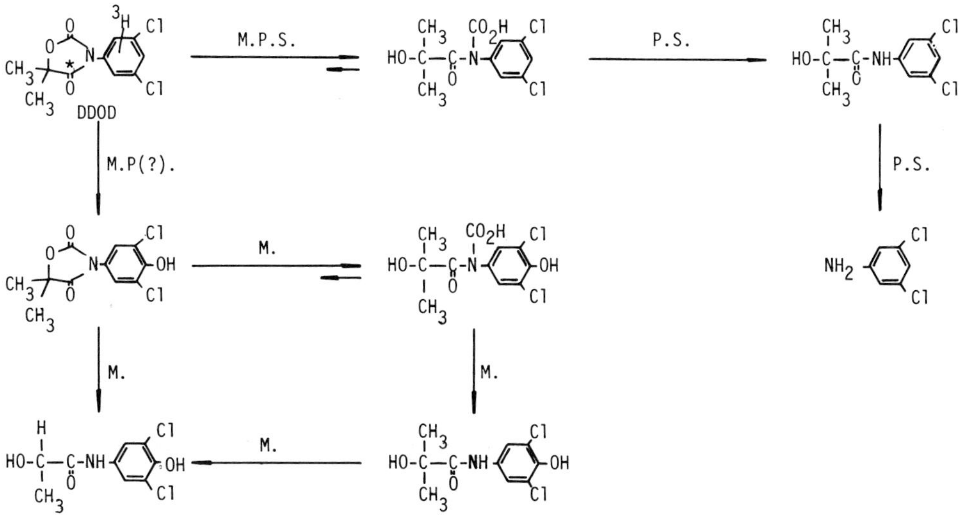

Mammal	Male Wistar rats[83]
Plant	Bean: *Phaseolus vulgaris;* Grape: Delaware[84]
Soil	84

METHAZOLE

Oxadiazolidine-3,5-dione is not stable enough to decompose to its idine phenylurea and benzimidazolinones by generating CO_2. Phenylurea is further metabolized by oxidation at the methyl group and ortho position of the phenyl ring. The LD_{50} of these metabolites in mice and rats are shown which indicates their low toxicity to mammals.

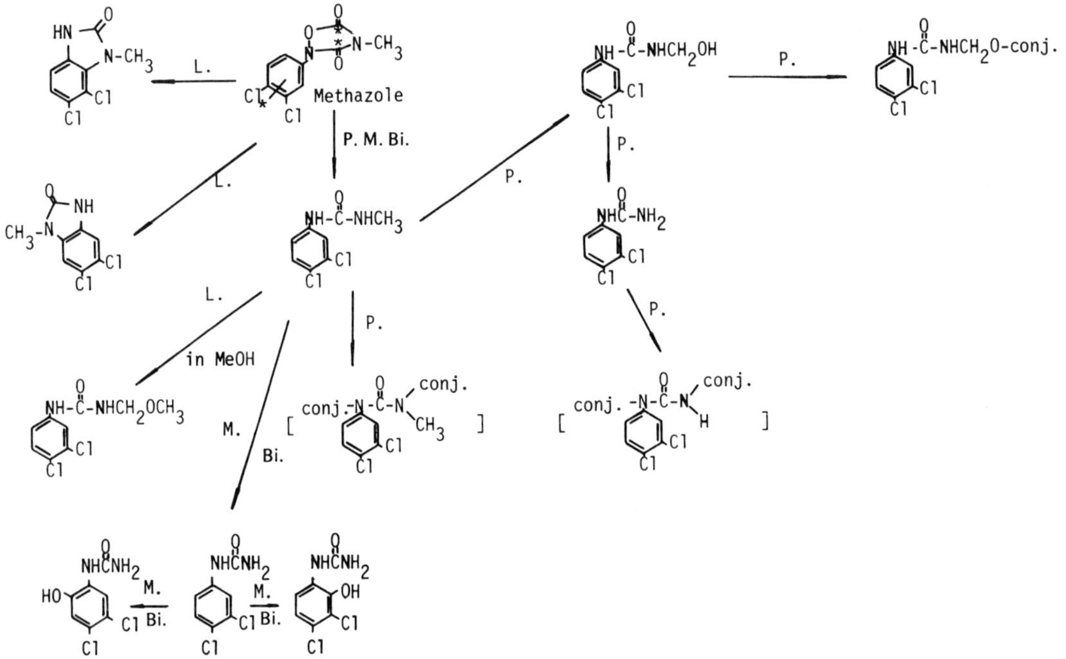

Light	Germicidal lamp; Sunlight[85]
Mammal	Holstein lactating cows[86]; Female albino rats[87]
Bird	White Leghorn laying hens[86]
Plant	Beans: *Phaseolus* sp.[87]

OXADIAZONE

Stability of the oxadiazone ring is altered by the location of heteroatoms in the ring. One oxygen and two nitrogen atoms are located at the 1, 2, and 4 positions in methazole and 1, 3, and 4 positions in oxadiazone. The methazole ring is less stable than that of oxadiazone, therefore all of the metabolites of oxadiazone have the oxadiazoline ring in their structures. The sequential demethylation reaction of the isopropyl group and oxidation reaction to alcohol and carboxylic acid of the *tert*-butyl group of oxadiazone are considered to be the main pathways.

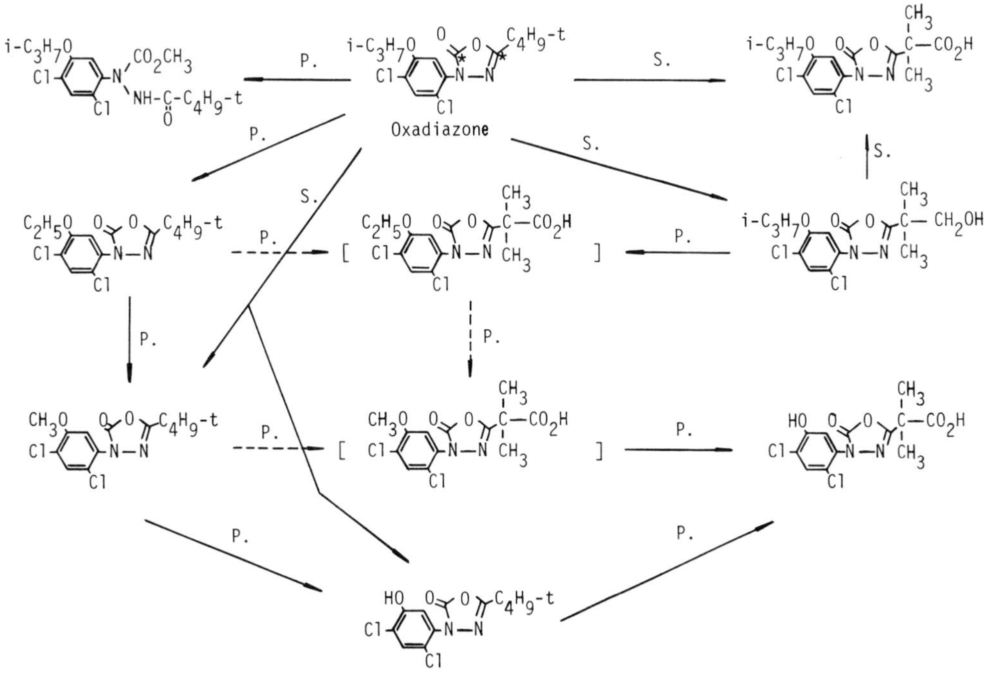

Plant	Rice seedlings: *Oryza sativa* L.v.Kinmaze or v. Nihonbare[88]
Soil	Metapeake loam soil; Monmouth fine sandy loam soil[89]

DIOXANE

There are three possible metabolic pathways of dioxane metabolism in the rat. The more toxic compound (ketodioxane) is formed (1) directly, (2) by hydroxylation at α-position and cleavage of the ring to aldehyde alcohol, or (3) by hydrolysis of the ring to diol and then by oxidation to acid alcohol intermediates.

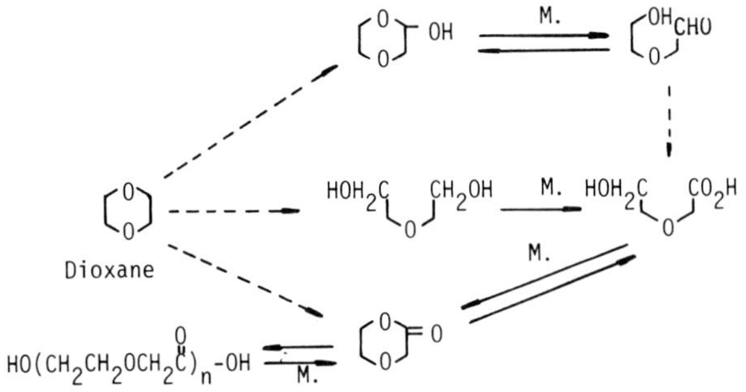

Mammal Male S-D rats[90]

TERBACIL

Oxidation of a methyl group and the *tert*-butyl group gives the oxidative products of terbacil, which are then cyclized to form an oxazolidine ring.

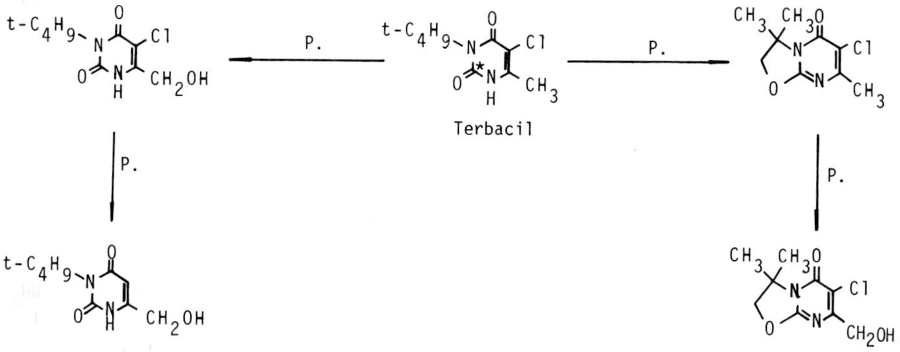

Plant Alfalfa[91]

NORFURAZONE (SAN-9789) AND SAN-6706

Ordinary N-dealkylations are usually observed in this 2-phenyl-5-(*N,N*-dimethylamino)-3-pyridazone herbicide. Monodemethylated SAN-9789 is more phytotoxic than SAN-6706.

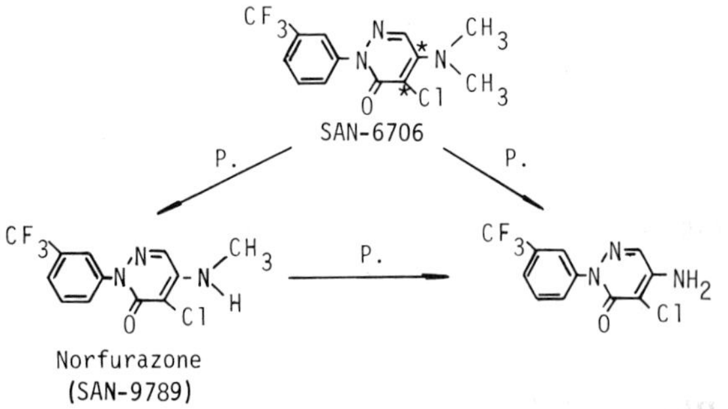

Plant Cotton: *Gossypium hirsuntum* L.'Coker 203'; Corn: *Zea mays* L.'WF 9'; Soybean: *Glycine* max. (L) Merr.'Lee'[92]

DIMETHIRIMOL

Metabolism of dimethirimol in plants follows a pattern similar to that in mammals, and three routes of metabolism, namely, N-dealkylation, hydroxylation of the butyl group, and conjugate formation have all been observed.

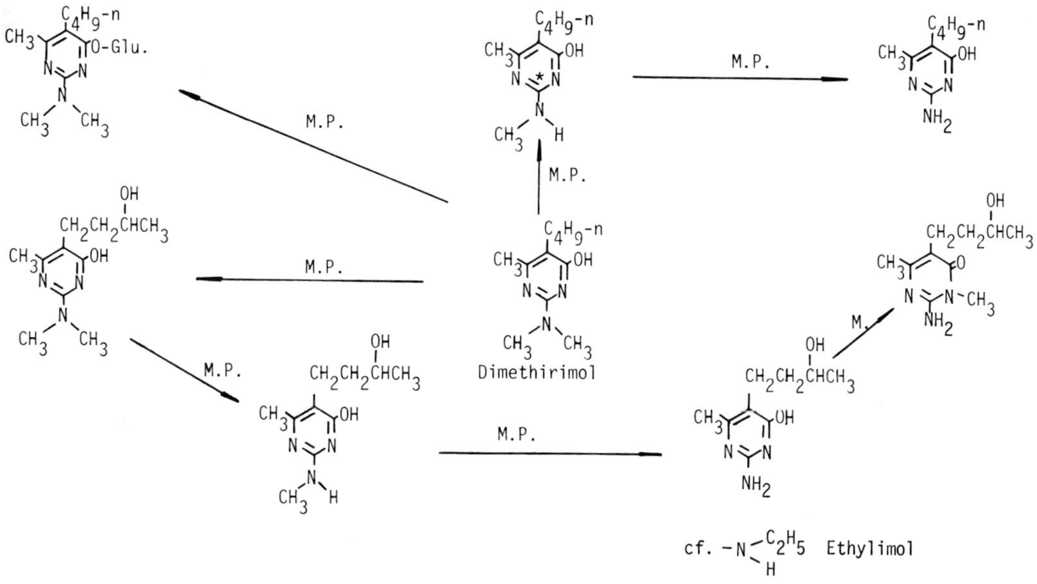

Mammal	Rats[93]
	Male Wistar rats (specific-pathogen free); Beagle dogs[94]
Plant	Cucumbers; Barley[93]

8

Imides

DSI [*N*-(3,5-DICHLOROPHENYL) SUCCINIMIDE]

Cleavage of the cyclic imide linkage is an important biotransformation step in the metabolism of DSI. Oxidation products bearing the succinic acid moiety have also been observed.

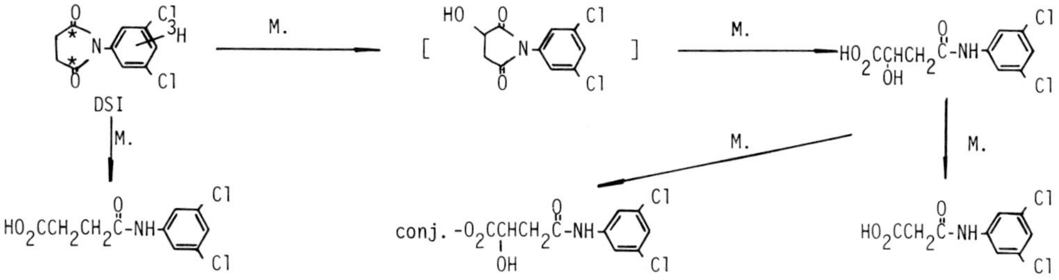

Mammal Female and male S-D rats; Male beagle dogs[95]

PROCYMIDON (SUMISCLEX)

Procymidon is rapidly metabolized via hydroxylation of a methyl group, subsequent oxidation to carboxylic acid in mammals, and cleavage of imide linkage in mammals, plants, and soil. Hydroxylation of the phynyl ring is characteristically observed in plants.

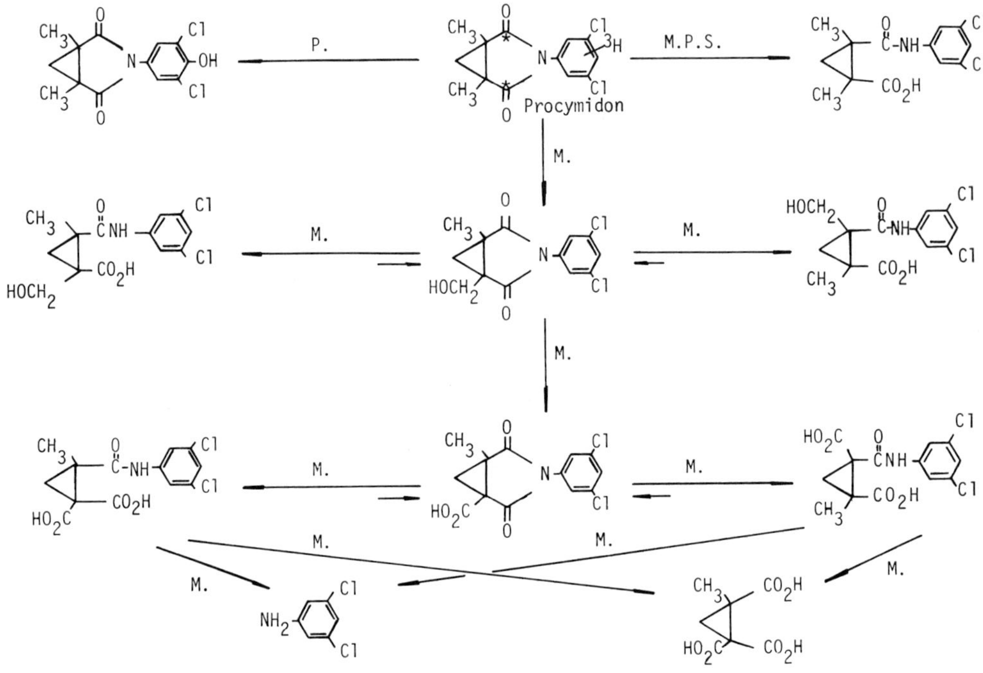

Mammal	Rats[96,97]
Plant	Cucumbers[96]
Soil	[96]

CAPTAN

Captan is easily hydrolyzed to 4-cyclohexene-1,2-dicarboximide via very unstable intermediates.

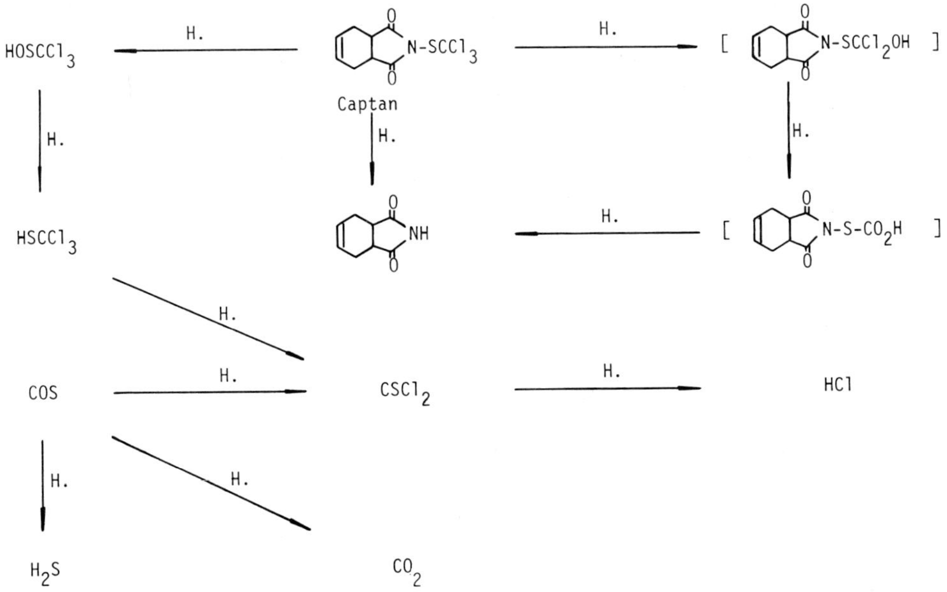

Hydrolysis 98

FLUOROIMIDE (SPARTCIDE, MK-23)

Fluoroimide, a dichloromaleimide derivative, is decomposed mainly by hydrolysis to *p*-fluorodichloromaleanilic acid, which is further degraded to acrylic acid and its amide by additional hydrolysis and decarboxylation. Two chlorine atoms are eliminated by reduction to maleimide, whose double bond is further reduced to give the corresponding succinimide.

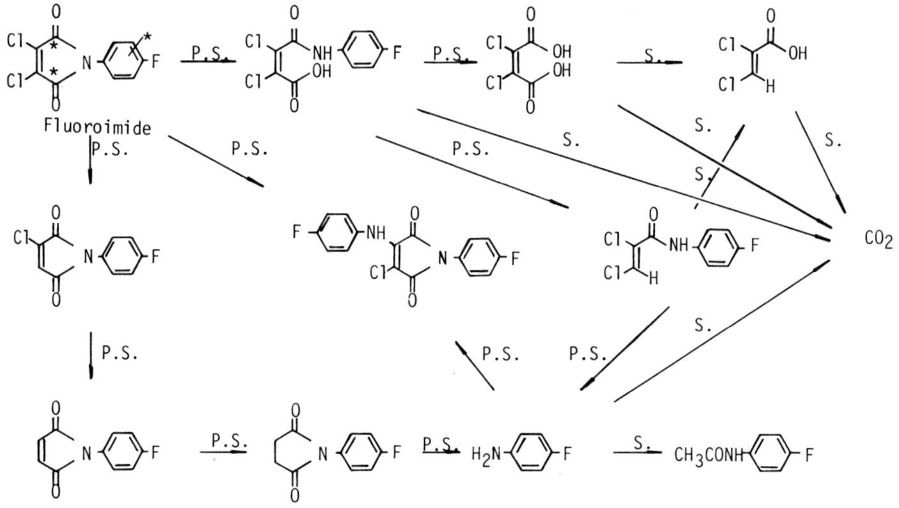

Plant Inu-apple tree: *Malus pruniforia*[99]

Soil Kumagaya clay; Tochigi clay loam soils[99]

9

Organochlorine Compounds

LINDANE

Newly identified phenolic metabolites of lindane in mammals are 2,3,4,6- and 2,3,4,5-tetrachloro-,2,4,6-trichloro- and 3, 4-dichlorophenol via 2,3,4,5,6-pentachloro-2-cyclohexen-1-ol.

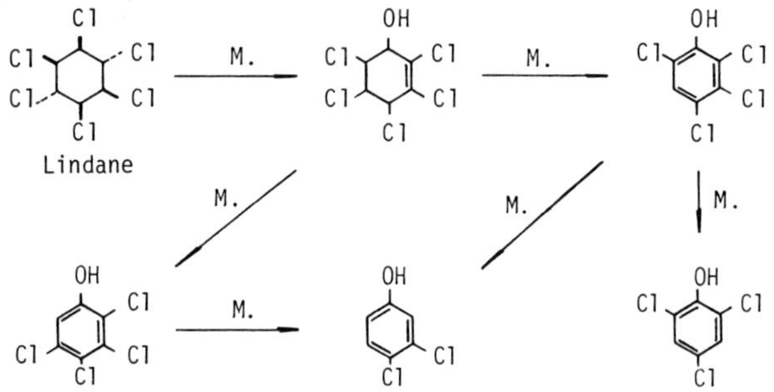

Mammal Weanling female S-D rats[100]

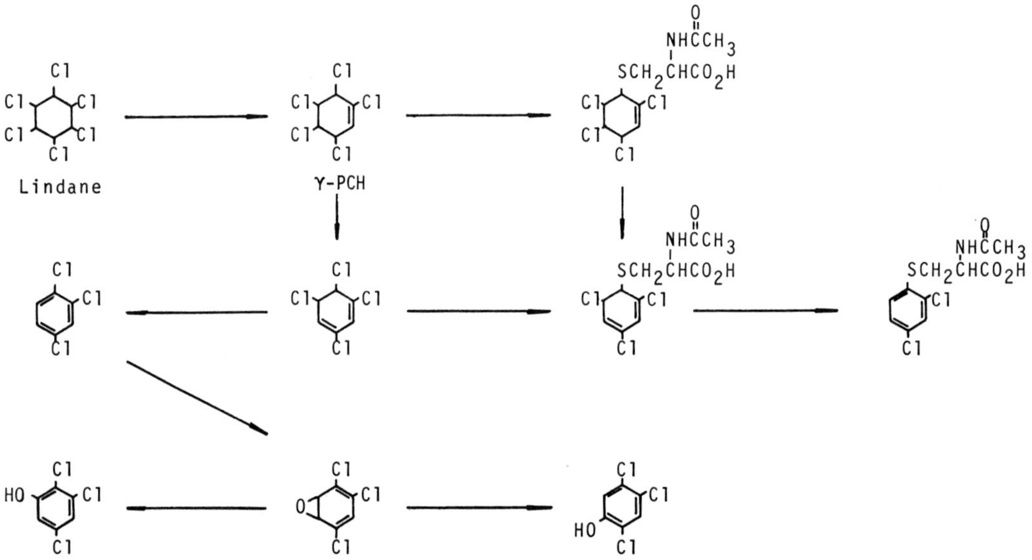

Previous lindane metabolic pathways proposed by Grover and Sims[100a] are shown above, where lindane is dehydrochlorinated to γ-PCH which is further metabolized to either 2,4-dichlorophenylmercapturic acid or 2,3-5- and 2,4-5-trichlorophenol.

POLYCHLORINATED NORBORNENES

In polychlorinated norbornenes, the 7-chlorine atoms are eliminated by reduction and/or replaced by a hydroxyl group to form *anti*-tetrachloronorbornen-7-ol. Decarbonation is observed when two chlorine atoms are substituted at the 7-position by emitting dichlorocarbene in bacterium.

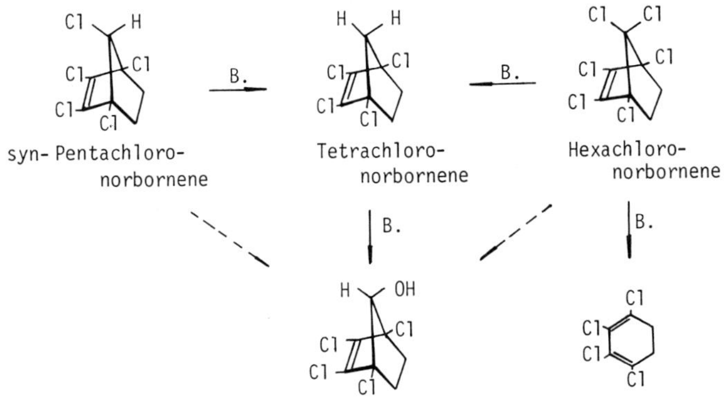

Bacteria *Clostridium butyricum*[101]

ENDOSULFAN

Conversion of endosulfan I to its sulfate occurs faster than that of endosulfan II. These three compounds are desulfurized to form a diol, which is then cyclized to give a γ-lactone as a final metabolite. Dechlorination does not occur.

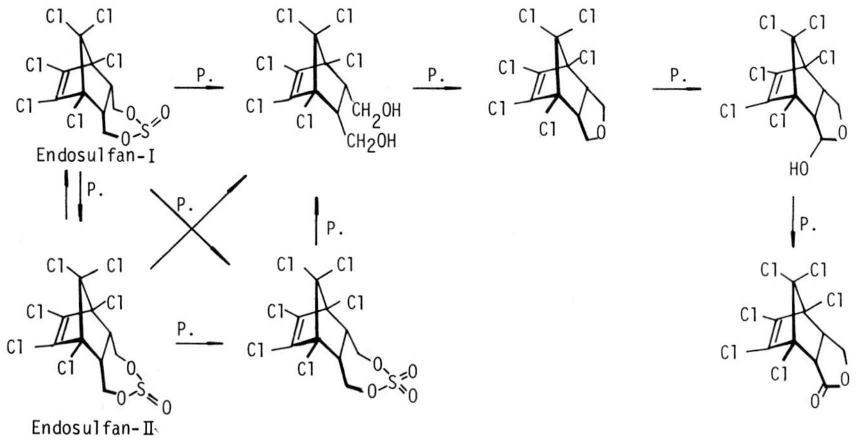

Plant Tobacco Coker 347 variety[102]

CIS-, TRANS-CHLORDANES

The major route of metabolism for both *cis*- and *trans*-chlordanes is via 1,2-dichlorochlordene and oxychlordene, which are further decomposed to two metabolites, 1-*exo*-hydroxy-2-chlorochlordene and 1-*exo*-hydroxy-2-*endo*-chloro-2,3-*exo*-epoxychlordene. These metabolites are not readily degraded further. *trans*-Chlordane is more readily metabolized through this route, but *cis*-chlordane is more readily degraded via a direct hydroxylation reaction to form 1-*exo*-hydroxydihydrochlordene and 1,2-*trans*-dihydroxydihydrochlordene.

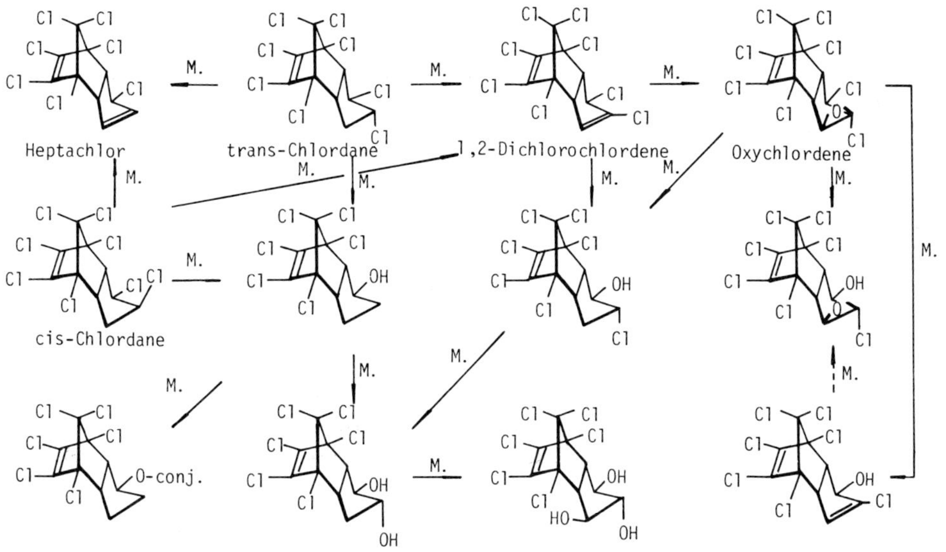

Mammal Male rats[103]

ENDRIN

Anti-12-hydroxylation of endrin gives the least toxic metabolite to resistant insect strains, although endrin itself accumulates more in susceptible strains of mice. The most toxic metabolite, 12-ketodieldrin, was not identified. Other hydroxylation reactions are observed to some extent, but all the chlorine atoms remain intact.

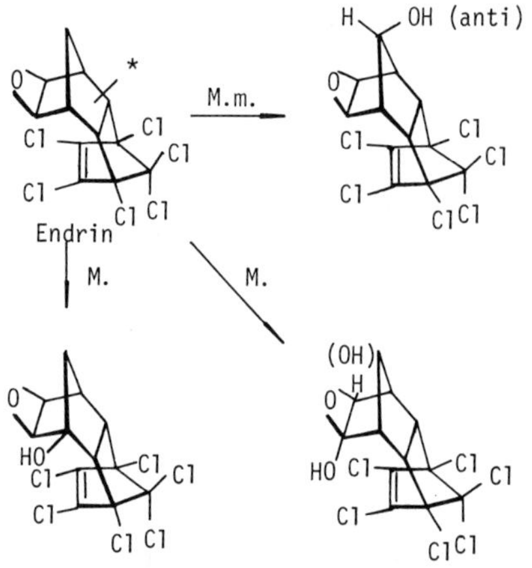

Mammal Susceptible and resistant strains of pine mice: *Microtus pitymus pinetorum*[104]

Microsome 104

ALDRIN

The double bond not substituted by chlorine atoms in aldrin is easily epoxidized to form the toxic metabolite dieldrin, which is then transformed to photodieldrin and hydroxylated dieldrin. Dihydroxylation via epoxidation at this double bond is characteristically observed in marine algae.

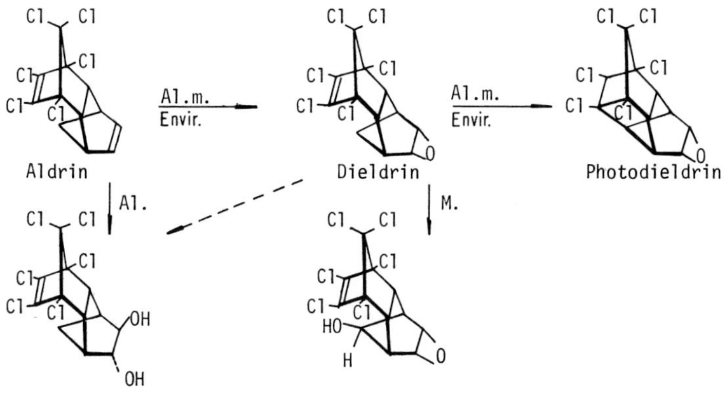

Mammal	Men (feces)[105]
Algae	Oceanic microorganisms = *Dunaliella* sp.; *Agmenellum quadraplicatum* (strain PR-6); Field collected algae containing water samples[106]
Environment	Aquatic food chain organisms = Algae: *Chlorella,* Diatoms; Protozoa: *Dinoflagellates;* Mixed-protozoa: *Hydra, Dugensia;* Leech: *Halobdella stangalish; Asellus,Gammarus,Daphnia,Cyclops* and *Aedes* larvae; Fresh water mussel: *Anodonta,* Snail: *Lymnaea*[107]
Microsome	Crayfish, Snail, Mussel tissue[107]

HCE (1,2,3,4,9,9-HEXACHLORO-*EXO*-5,6-EPOXY-1,4,4a,5,6,7,8,8a-OCTAHYDRO-1,4-METHANONAPHTHALENE)

As in the case of dieldrin, the epoxide of HCE is hydrated to give the corresponding diol and is hydroxylated at the alkyl chain not substituted by chlorine atoms to give mono- and dihydroxylated HCE. Dechlorination does not occur in this metabolic pathway.

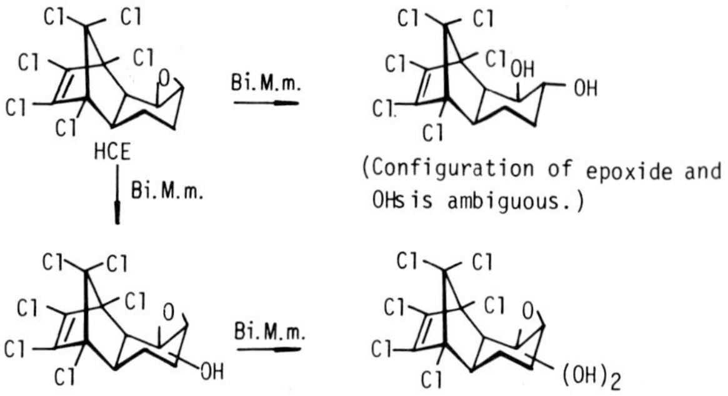

Mammal — Adult male Wistar rats; Old English rabbit

Bird — Pigeons: *Columba livia;* Japanese quail: *Coturnix coturnix japonica*

Microsome — Jackdaws (*Corvus monedula*), and rooks (*Corvus frugilegus*) liver microsomes[108]

DIHYDROCHLORDENEDICARBOXYLIC ACID

Dihydrochlordendicarboxylic acid, a rather highly oxidized insecticide, undergoes dechlorination to give a tetrachlorinated compound in microsomes.

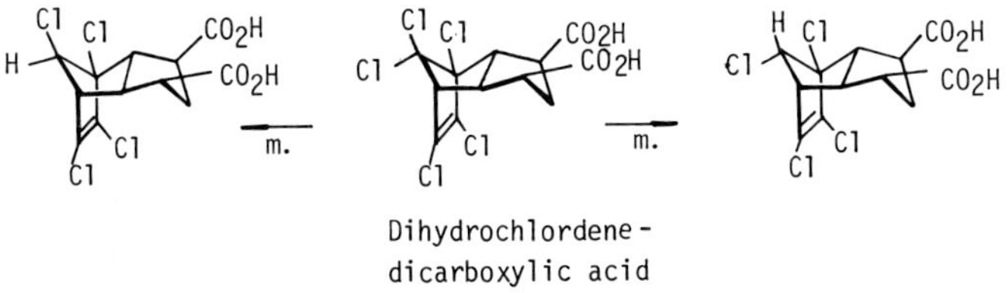

Microsome Rat liver microsomes[109]

10

Oxime Carbamates

THIOFANOX (DS-15647)

The major route of metabolism appears to be the oxidation of thiofanox to its sulfoxide and the subsequent oxidation to its sulfone. The rate of the initial reaction is decisively more rapid than that of the second, particularly in plants. Hydrolysis of carbamoyl moieties to their resulting oximes is also observed. LD_{50} values (mg/kg) of thiofanox and related compounds in rats (oral) and rabbits are listed. Sulfonyl oxime carbamate is most toxic.[111]

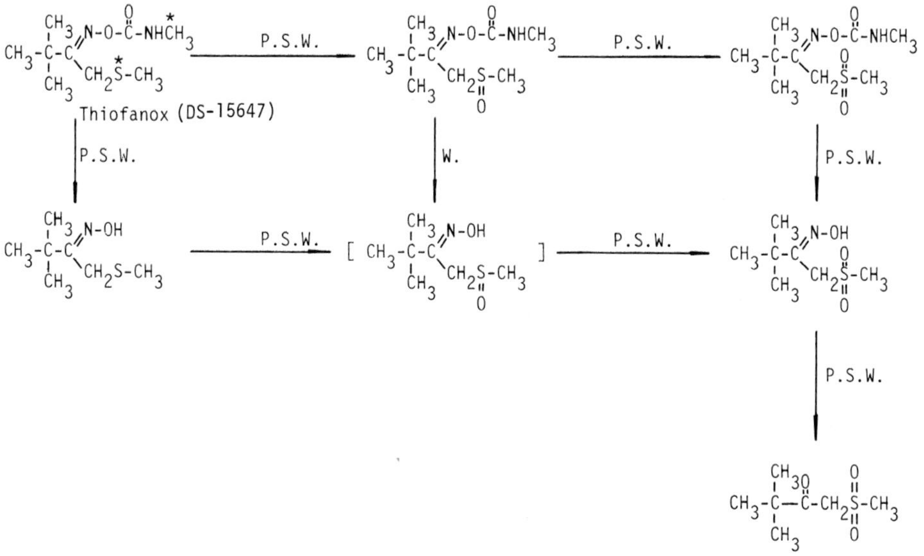

Plant	Cotton: Deltapine smoothleaf variety[110] Potatoes, Sugarbeets, Foliage, Cotton seeds[111] Cotton: *Gossypium hirsuntum* L.var. Stoneville 213[112]
Soil	Air-dried 12 mesh soil[111]
Water	[111] Aqueous solutions[113]

ALDICARB

Sulfone forms of aldicarb metabolites are predominant. Cyanosulfone is another major metabolite identified.

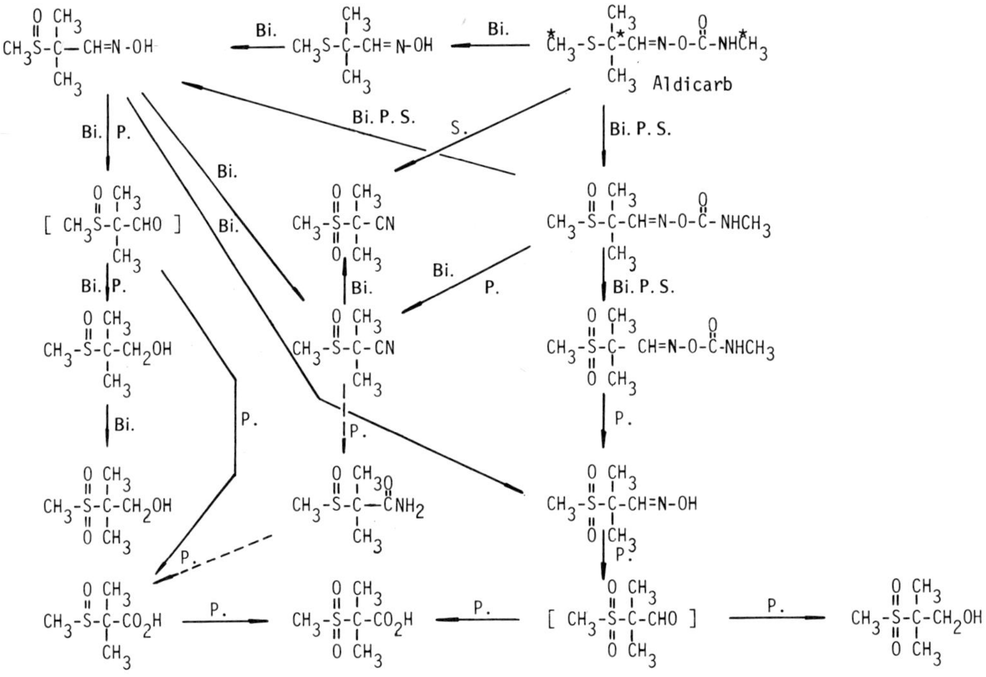

Plant	Cotton: *Gossypium hirsuntum* L.var. Coker 100[114]
Soil	Sandy loam soil (pH 6.0)[115] Lufkin fine sandy loam (pH 7.1); Lakeland fine sand (pH 4.6); Norfolk sandy loam (pH 4.8)[116]
Bird	White Leghorn laying hens[117]

OXAMYL

Sulfur atoms directly bound to imino carbon atom are not oxidized to sulfoxide or sulfone in microsomes except when the sulfur atom binds to a carbon atom other than the imino carbon in some oxime carbamates. Dealkylation, hydrolysis, and formation of a cyano group are included in this metabolic pathway.

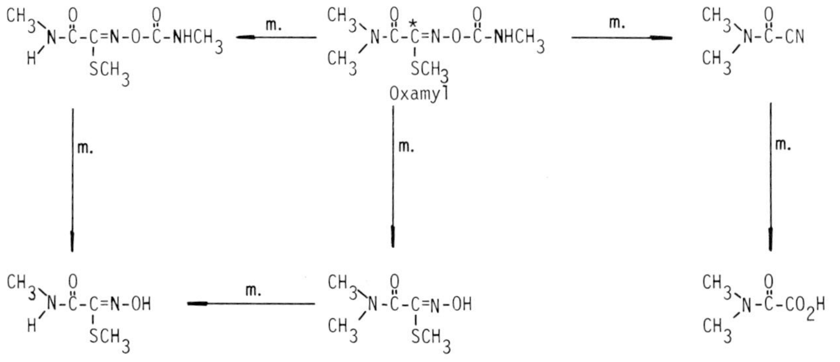

Microsome	Liver microsomes of Charles River-CD rats[118]

TRIPATE

One of the terminal metabolites is bisulfone in the 1,3-dithiolane moiety of tripate, which possesses an oxime carbamoyl group in the 2-position. 2-Cyano-1,3-disulfolane and 2-carboxy-1,3-dithiolane are the other terminal products in plants.

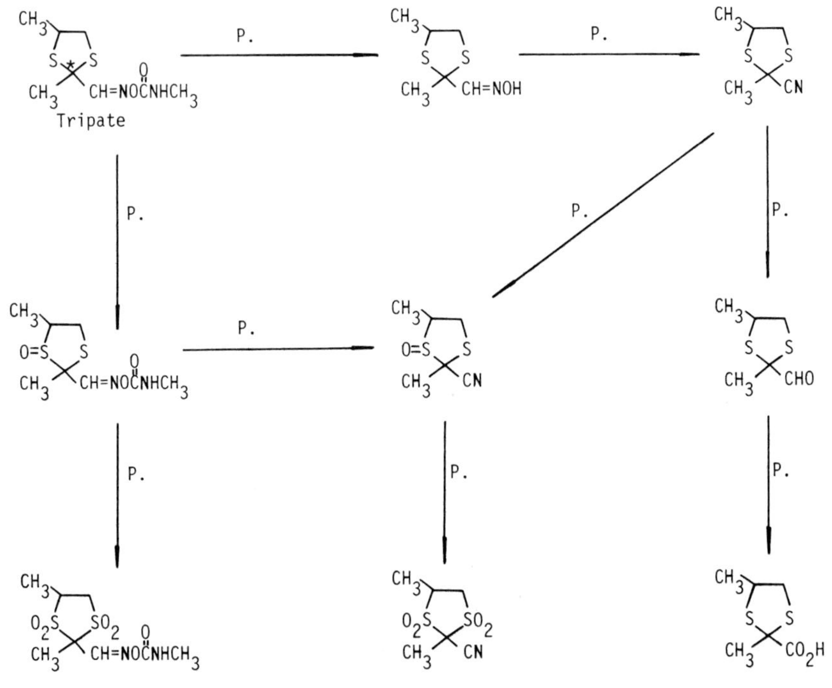

Plant Tobacco: *Nicotiana tabacum* var. Coker 319[119]

11

Phenoxyacetic Acids

4-CPA (4-CHLOROPHENOXYACETIC ACID)

The *p*-substituted chlorine atom of 4-CPA is displaced by a hydroxyl group during hydrolysis and by a hydrogen during reduction, respectively. The resulting *p*-chlorophenoxyformaldehyde is decomposed to *p*-chlorophenol and then to *p*-hydroquinone by hydrolysis in light.

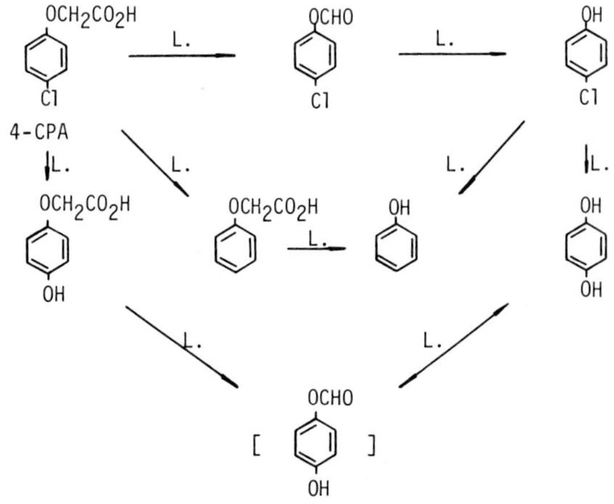

Light UV light[120]

2,4-D (2,4-DICHLOROPHENOXYACETIC ACID)

Various phenyl ring-hydroxylated metabolites of 2,4-D have been observed in plants and plant callus. NIH-shifted metabolites are examples of these metabolites. 2,4-Dichlorophenol is the only degradative compound that retains the phenyl ring of 2,4-D.

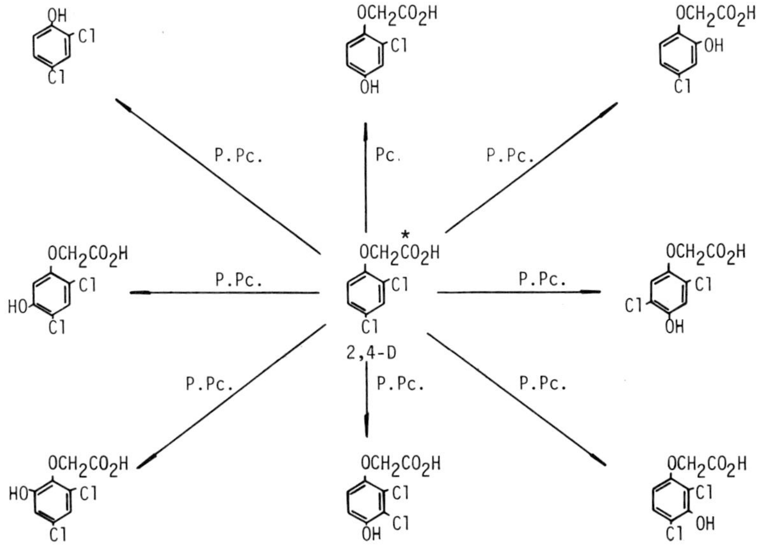

Plant Callus — Jack-bean (*Canavalia ensiformis*) pod callus; Sweet corn (*Zea mays*) endosperm callus; Tobacco (*Nicotiana tabacum*) pith callus; Carrot (*Daucus carota* var. *sativa*) pith callus; Sunflower (*Helianthus annus*) pith callus[121]

Plant — Rice: *Oryza sativa* var. Starbonnet[122]
Soybeans: *Glycine max.* (L.) Merrill var. *Acme;* Corn: *Zea mays* (L.) [Su-1][123]

2,4-DB (2,4-DICHLOROPHENOXYBUTYRIC ACID)

A typical β-oxidation product (2,4-D) is observed via phenoxybutenoic acid in 2,4-DB metabolism by plants. The decanoic acid derivative of 2,4-dichlorophenol is an interesting metabolite.

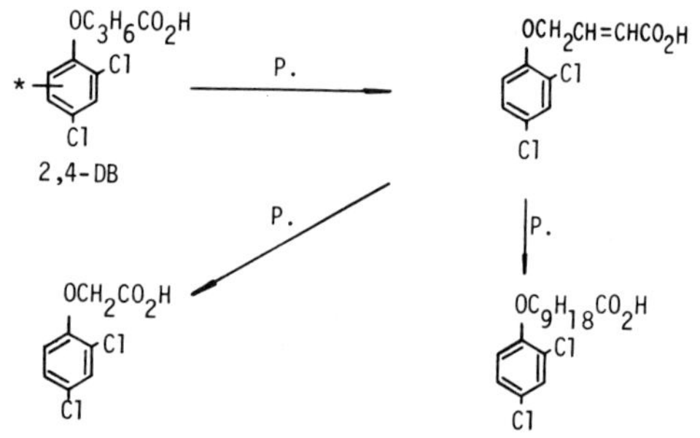

Plant Soybean: *Glycine max.* L. Merr. var. Lee; Cocklebur: *Xanthium* sp.[124]

PHENOTHIOL

Phenothiol, which is an ethyl thiol ester of MCP, is hydrolyzed to carboxylic acid (MCP) which is further decomposed to *p*-chlorophenol. Hydroxylation of the methyl group and at the 6-position of the phenyl ring of MCP comprise the secondary metabolic step of phenothiol in plants.

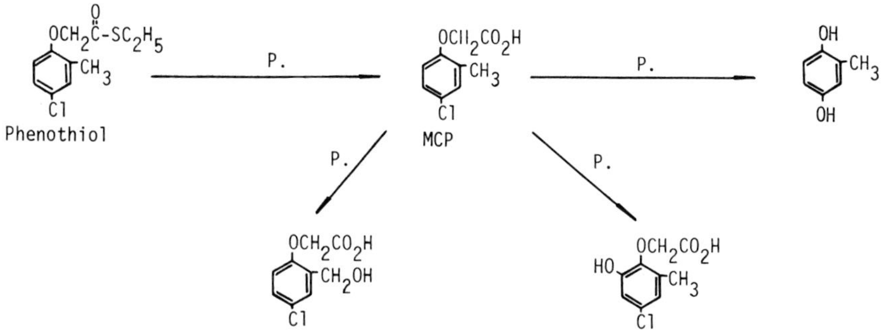

Plant 5

2,4,5-T
(2,4,5-TRICHLOROPHENOXYACETIC ACID)

Removal of the side chain of 2,4,5-T gives 2,4,5-trichlorophenol; this is followed by replacement of the ring chlorines by a hydroxyl group or a hydrogen. When the ring chlorine at the 2-position is replaced by a hydroxyl group in 2,4,5-T, the 1,4-dioxane ring is formed by dehydration between carboxyl and hydroxyl groups. Toxic 2,3,7,8-tetrachlorobenzo-*p*-dioxin (TCDD), which might be formed from 2,4,5-trichlorophenol by initial phototonucleophilic displacement of the *o*-chlorine by the trichlorophenoxide ion followed by ring-closing displacement, is not detected due to the instability of TCDD to light.

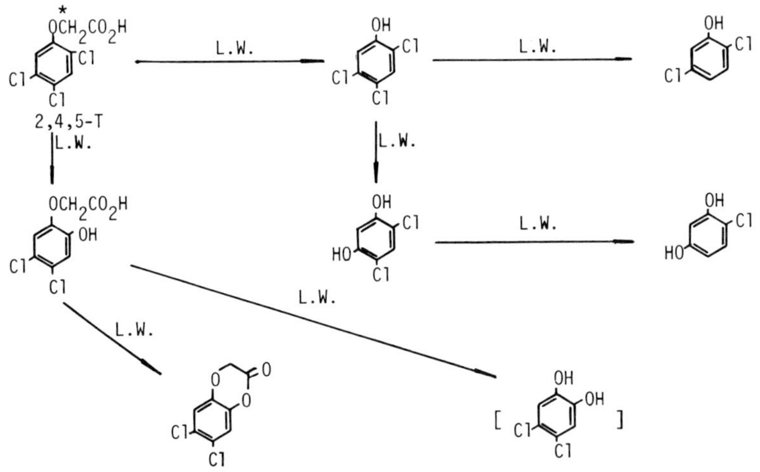

Light UV light, in indoor and sunlight in outdoor conditions[125]

Water Distilled water[125]

DICHLORFOP-METHYL [HOE-23408(OH)]

Dichlorfop-methyl undergoes extensive hydrolysis to its corresponding carboxylic acid in plants and in soil. It is then decarboxylated to phenyl ethyl ether and transformed to the phenol in soil. The phenol may then undergo complexing or binding to soil and/or be subjected to further degradative processes involving ring fission to give such products as *m*-dichlorobenzene or 2,4-dichlorophenol. Ring hydroxylation at the 5-position of the 2,4-dichlorophenol moiety was observed but not at the other positions of the phenyl rings in plants and soil.

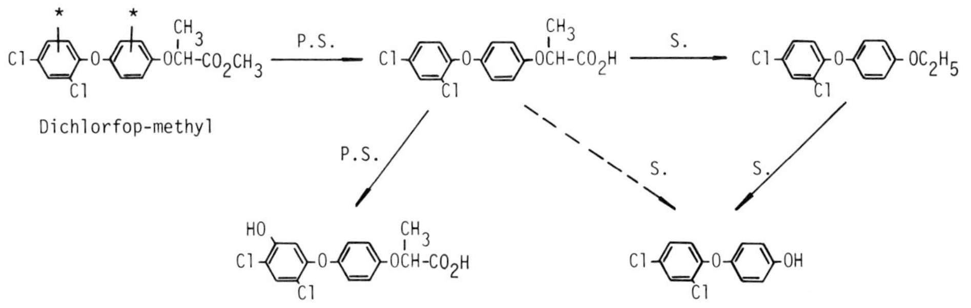

Plant — Summer wheat: *Triticum aestivum,* Colibri strain[126]

Soil — Jameson sandy loam (pH 7.5); Melfort silty clay (pH 5.2); Regina heavy clay (pH 7.1)[127]

12

Phenyl Ring Fused Five-Membered Heterocyclic Compounds

FTHALIDE (3,4,5,6-TETRACHLOROPHTHALIDE)

To investigate the phytotoxicity of the degradative products, fthalide was added to compost. This polychlorinated fungicide degrades via three routes, the first of which is oxidation to phthalic acid, the second is stepwise dechlorination, and the third is replacement of the ring chlorine by the SMe group. These degradative compounds are finally metabolized to phthalic acids. Some of these metabolites show slightly higher phytotoxicity than fthalide, but in amounts too small to cause damage to vegetables.

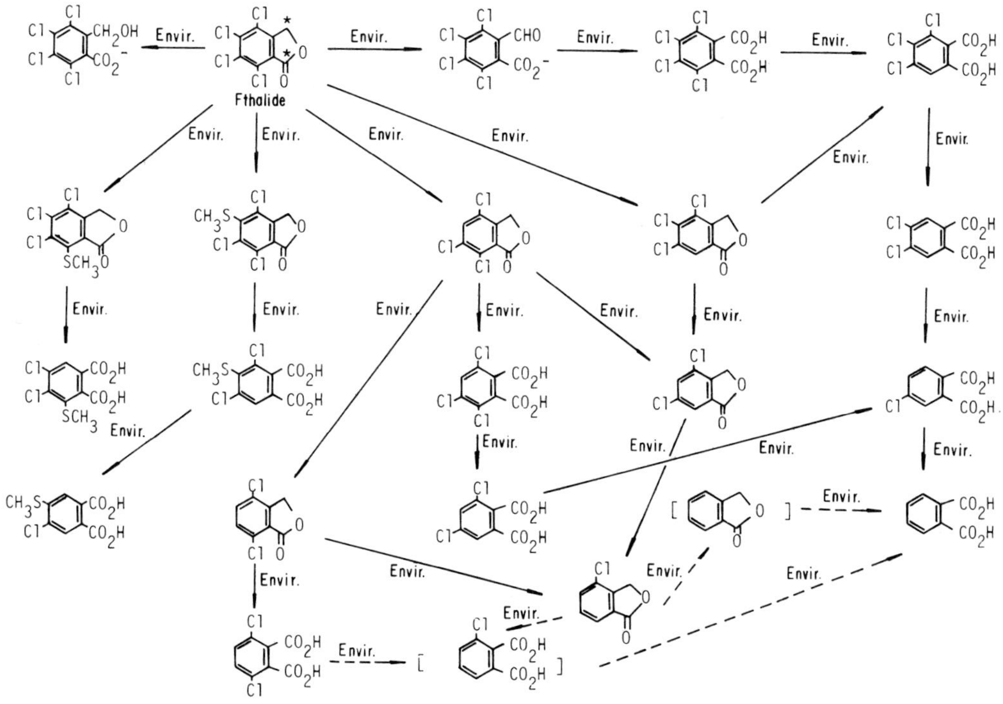

Environment Compost[128]

THIOPHANATE METHYL AND THIOUREIDOBENZENES

These *o*-phenylenediamine derivatives are generally able to transform to benzimidazole analogs in the environment. Their phenyl rings are hydroxylated at selective positions in fungi and mammals. N-Substituted groups (R_2) of these *o*-phenylenediamines, which may have many kinds of groups (see references) are metabolized stepwise in mammals.

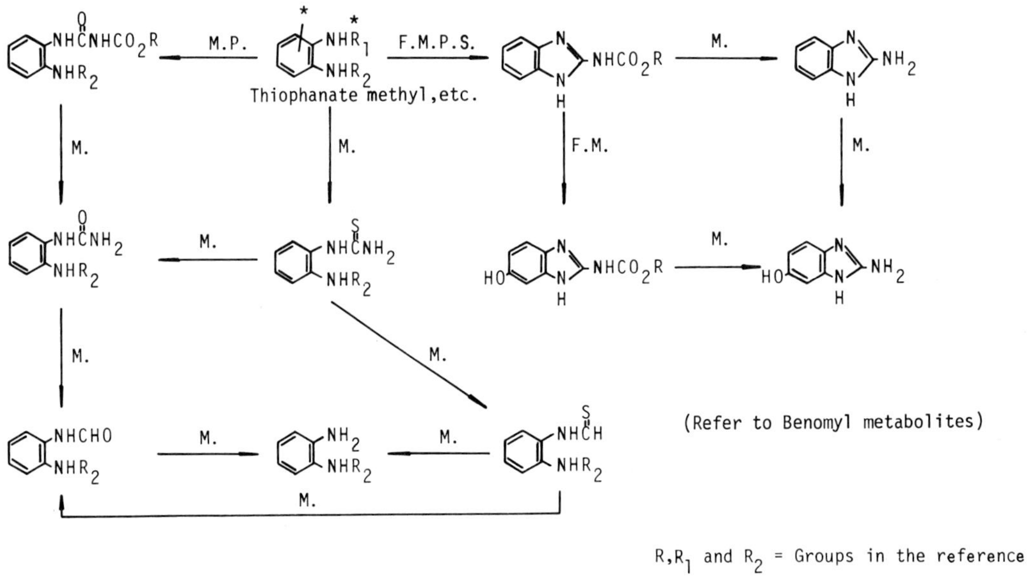

Plant	Bean: *Phaseolus vulgaris* L.[129] Apple: *Malus pumila* M. (Kokko variety); Grapes: *Vitis uiniferalabrusca* L. (Delaware variety)[130]
Fungi	*Pellicularia sasakii* (Shirai); *Alternalia mali* Roberts[131]
Soil	Barnes sandy loam (pH 7.4); Fargo silty loam (pH 7.7); Towner loamy fine sand (pH 7.2)[132]
Mammal	Male mice; Romney-Southdown cross wether[133]

BENOMYL

Benomyl is metabolized to various compounds including ring-opening products of the benzimidazole ring to produce *o*-phenylenediamine, *o*-aminobenzonitrile, methoxycarbonylguanidine, methoxycarbonylurea, and 2-amino-6-hydroxyphenylurea in light and plants. In mammals metabolites include hydroxylated products at the 4,5, and 6 positions of the phenyl ring of benzimidazole.

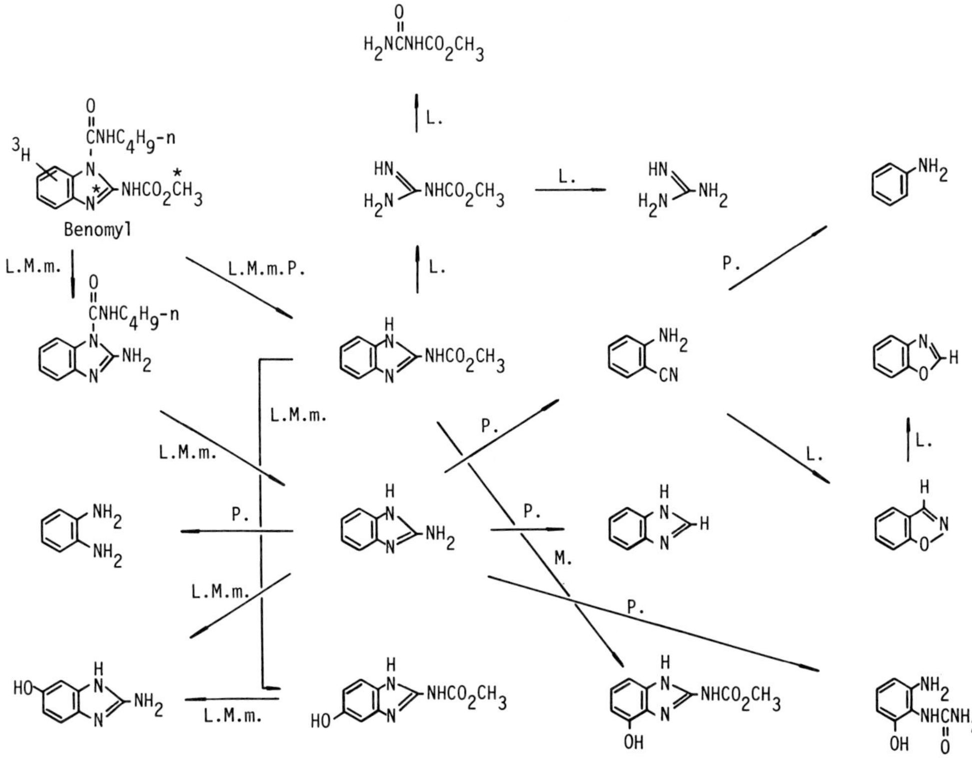

Mammal	Cow milk; Cream; Feces; Urine; Tissues[134] Male Charles River-CD rat; Male beagle dog; Guernsey dairy cows[135] Male mice; Male New Zealand white rabbits; Sheep: Romney-Southdown cross wethers[136]

Light	Photolysis[136]
	Photolysis on TLC plate and in acetone solution; Sunlight irradiation on corn plant (*Zea mays*) leaves[137]
Microsome	Liver enzyme preparation (mouse kidney, brain, finely minced intestine and blood; Sheep rumen fluid)[136]
Plant	Melon: *Cucumis melo* L. 'Noir des Carmes'[138]
	Melon: *Cucumis melo*[139]

PARBENDAZOLE

The *n*-butyl group of parbendazole, substituted at the 5-position of the benzimidazole ring, a position easily hydroxylated in the environment, mainly undergoes metabolic attack to form monohydroxylated, dihydroxylated, and carboxylated products in addition to a ring-hydroxylated metabolite of parbendazole in mammals.

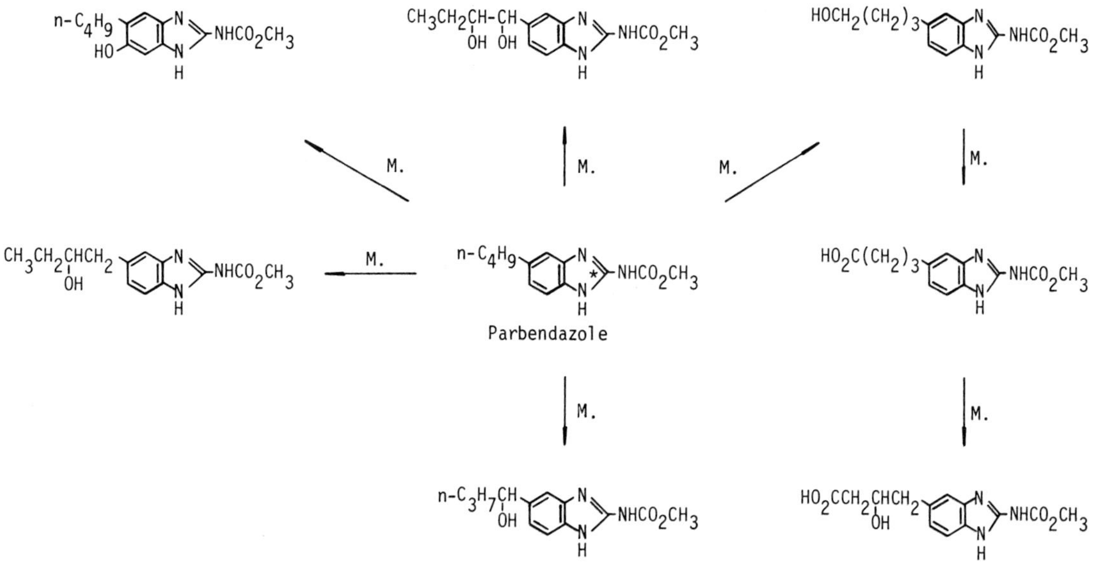

Mammal Sheep (lamb)[140]

METHABENZTHIAZURON

Methabenzthiazuron possesses *N,N'*-dimethylureido and benzimidazole moieties. Demethylation and the resulting deamidation in ureido groups and hydroxylation in the phenyl ring occur, but none of the ring-opening products of benzimidazole are observed.

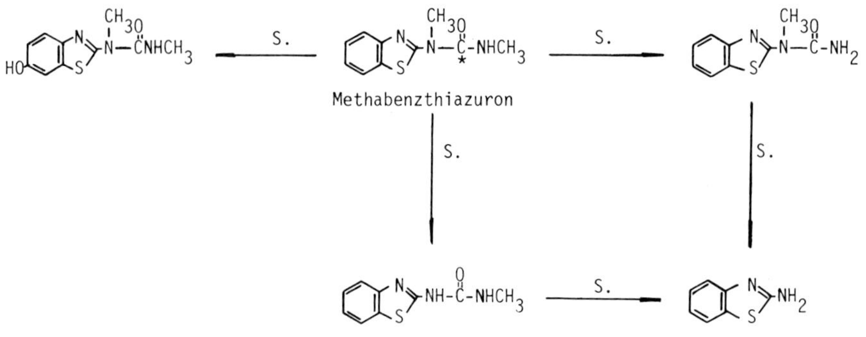

Soil Degraded loess soil (pH 6.3)[141]

ORYZEMATE

Saccharin which is a final stable degradation product of oryzemate is converted from allyl-*o*-sulfaminobenzoate formed by ring-opening hydrolysis of oryzemate in plants.

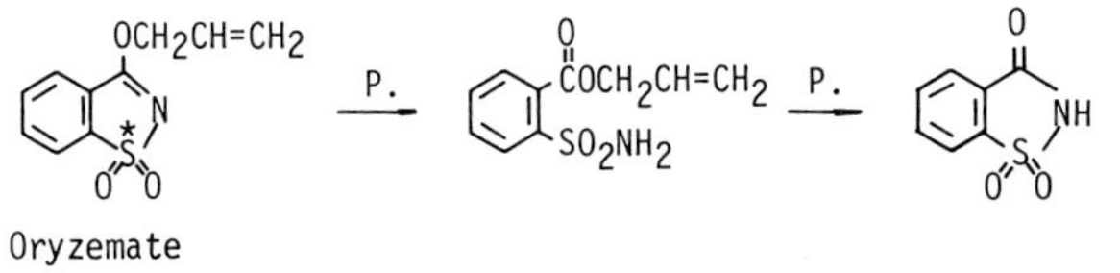

Plant Rice: *Oryza sativa* 'Satohikari'[142]

13

Phenyl (Aryl) Carbamates

MTMC (TSUMACIDE)

In mammals, MTMC is mainly metabolized by oxidative reactions at the meta methyl group and the para-position of the phenyl ring. Hydrolysis of the ester linkage occurs only to a minor extent. Oxidation of the meta-methyl group, the *N*-methyl group, and the para-position of the phenyl ring are observed with MTMC in insects and plants.

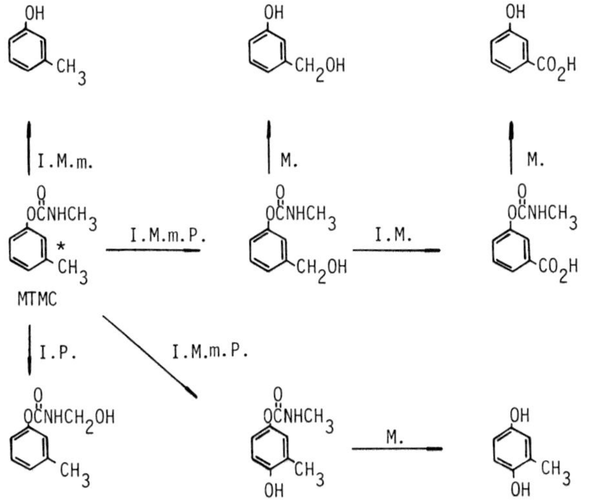

Insect	Female houseflies (Takatsuki strain)[143]
Mammal	Female and male S-D strain rats[143]
Microsome	Mouse, rat, and rabbit liver homogenate + NADPH[143]
Plant	Bean: *Phaseolus vulgaris*[143]

MIPC (MIPCIN)

Hydrolysis of the ester linkage in various carbamates must be considered one of the degradative pathways in many metabolic systems. This reaction is not an exception in the metabolism of MIPC. The isopropyl group substituted at the 2-position of the phenyl ring of MIPC is susceptible to oxidation to form two monohydroxylated compounds at the benzyl position and one of the methyl groups, the latter of which is finally oxidized to carboxylic acid. An N-demethylated product resulted from hydroxylation of the *N*-methyl group and another hydroxylated product at the 4-position of the phenyl ring are also observed.

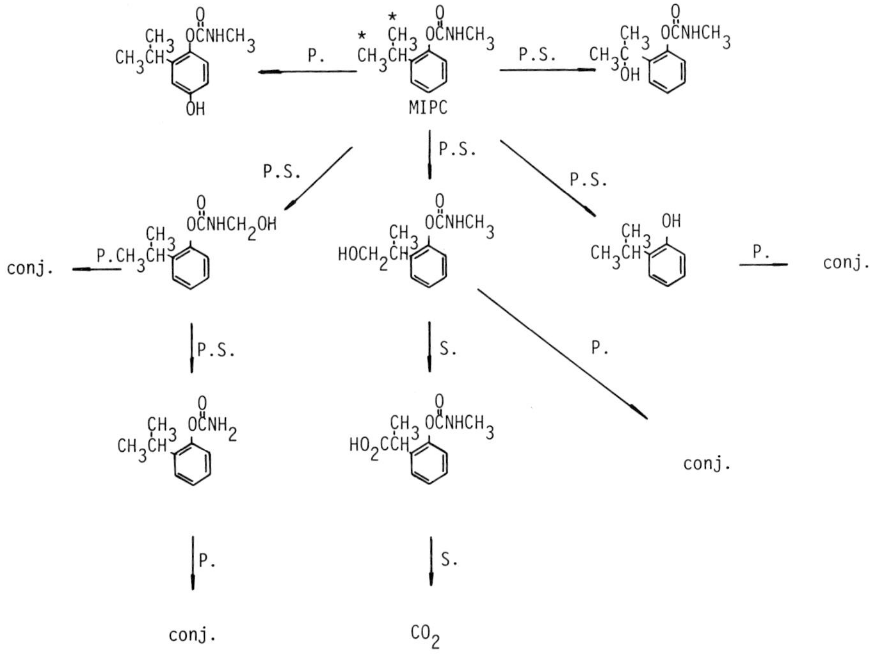

Plant Rice: Sasanishiki variety[144]

Soil Paddy field soils: Tochigi volcanic ash soil (pH 6.4), Kochi mineral soil; Upland soils: Tochigi volcanic ash soil (pH 6.4), Kochi mineral soil[144]

BPMC (BASSA)

The *sec*-butyl group at the 2-position of the phenyl ring undergoes oxidative degradation, which results in oxidation products at all carbon atoms of the butyl group. Hydrolysis and N-demethylation are also observed in its metabolic pathways.

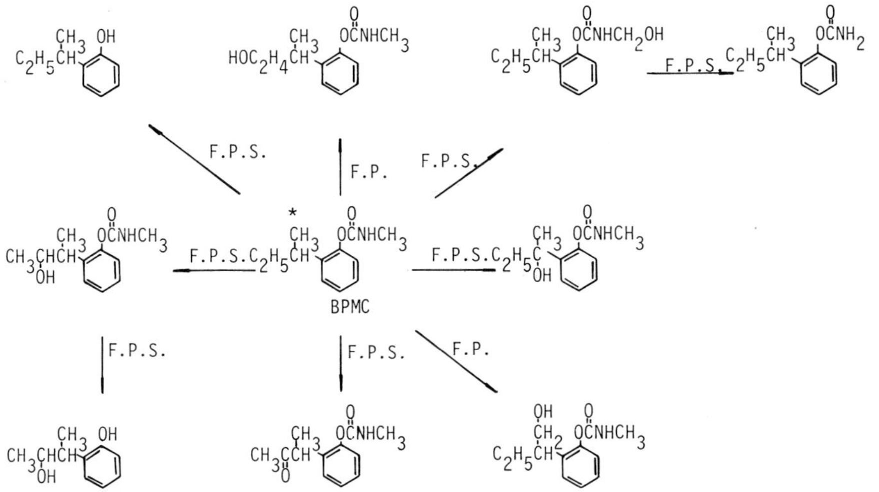

Plant	Rice: 'Jukkoku' and 'Sasanishiki'[145]
Soil	Paddy field soils: Tochigi volcanic ash soil (pH 6.4), Kochi mineral soil; Upland soils: Tochigi volcanic ash soil (pH 6.4), Kochi mineral soil[145]
Fungi	*Aspergillus niger* var. Tieghem[146] *Fusarium oxysporum, Penicillium funiculosum, Cladosporium cladosporioides, Coniothyrium* sp.[147] *Cladosporium cladosporioides*[148]

MTBC (*M-TERT*-BUTYL PHENYL *N*-METHYL CARBAMATE)

Only one of the methyl groups of the *tert*-butyl side chain substituted at the 3-position of the phenyl ring of MTBC is hydroxylated, in addition to N-hydroxymethylation and the resulting N-demethylation and hydrolysis to the corresponding phenol in insects and mammals.

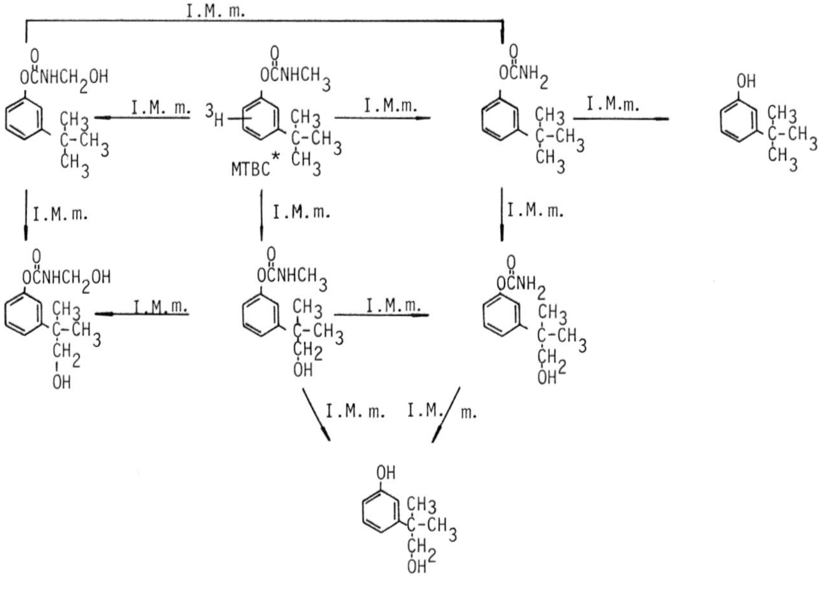

Insect Houseflies: *Musca domestica;* Blowflies: *Lucia scricata;* Grass grubs: *Costelytra zealandica;* Bees: Workers from *Apis mellifera* colony; Mealworms: *Tenebrio* sp.[149]

Mammal Male mice[149]

Microsome Mouse liver microsomes and enzymes from houseflies or blowflies[149]

BUX

Bux, which is a 3-to-1 mixture of *m*-(1-methylbutyl)phenyl methyl carbamate and *m*-(1-ethylpropyl)phenyl methyl carbamate is readily metabolized at the carbamate ester linkage by soil organisms. Only the benzyl position of both of the carbamates is hydroxylated to give the identified metabolites.

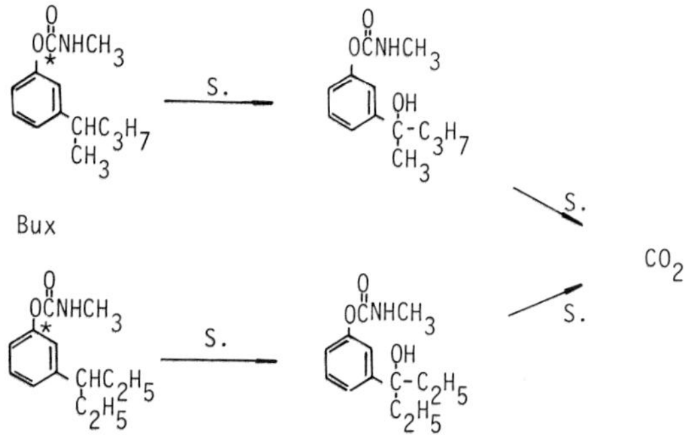

Soil Iowa silt soil[150]

LANDRIN-1 AND -2

The methyl groups of landrin-1 are oxidized stepwise to corresponding alcohols, aldehydes, and carboxylic acids, all of which undergo hydrolysis to give the phenols as shown. The 2-position of the phenyl ring is hydroxylated in insects and microsomes.

The methyl group at the 2-position of the phenyl ring of landrin-2 is resistant to degradation because it is sterically hindered by the adjacent carbamoyl and methyl groups. Also it may be mentioned that hydroxylation at the 6-position of the phenyl ring is more difficult than at 4-position where the proton is less hindered.

Landrin-1

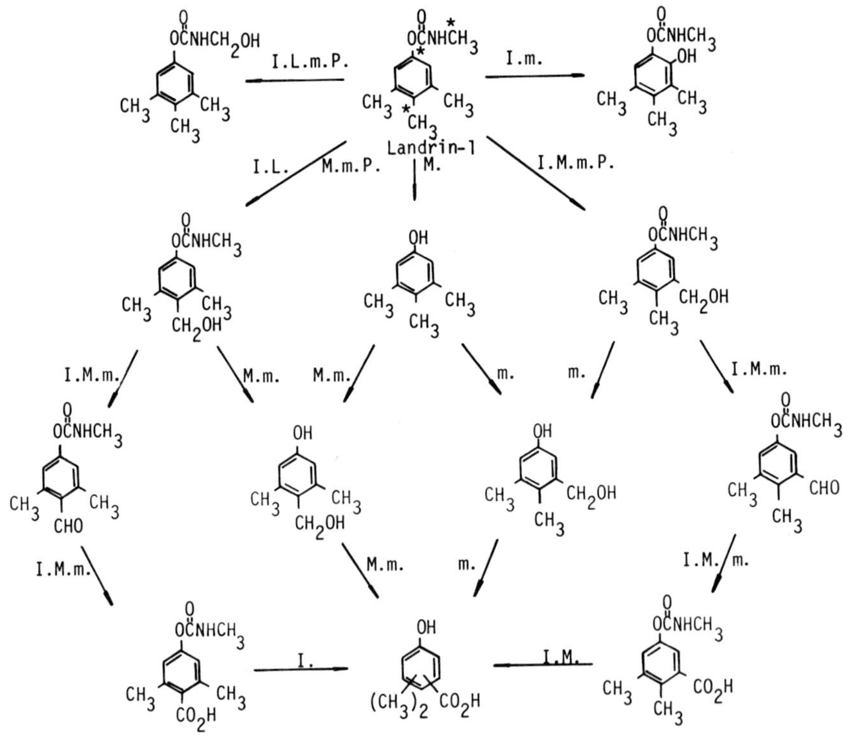

Landrin-2

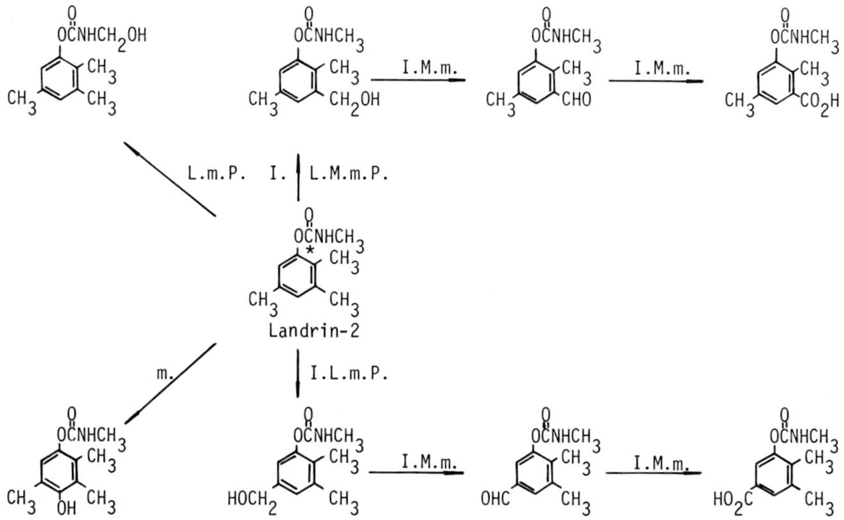

Insect	Adult female houseflies: SCR-susceptible strain[151]
Light	Sunlight on leaves of growing bean plant (Pinto variety)[151]
Mammal	Male albino mice: Swiss-Webster strain[157]
Microsome	Mouse liver microsomes and housefly (*Musca domestica* L.,R_{Baygon} strain) abdomens[151]
Plant	Snapbean: Contender variety[151]

BANOL

Banol is metabolized by direct conjugation in mammals and is hydrolyzed to phenol followed by conjugation in mammals and microsomes. The methyl groups in banol are not oxidized probably because of the chlorine substituted at the 2-position of the phenyl ring.

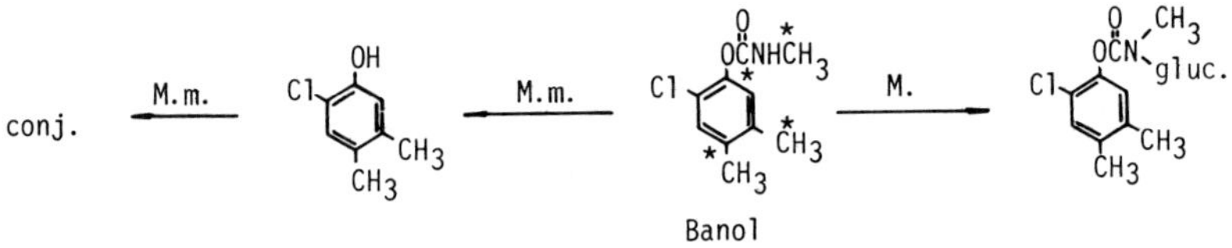

Microsome	Human embryonic lung cell culture (L-132 strain)[152]
Mammal	Rat[153]

BPBSMC AND BPMC (*M-SEC*-BUTYL PHENYL *N*-METHYL CARBAMATE)

Oxidation of the sulfur atom of BPBSMC to sulfone is observed only in photolysis. Although the fate of thiophenol which is released by hydrolysis in many systems is not determined, the residual BPMC is oxidatively metabolized on the *m-sec*-butyl group to give four alcohols, one ketone, and two carboxylic acids hydrolyzed to the corresponding phenols.

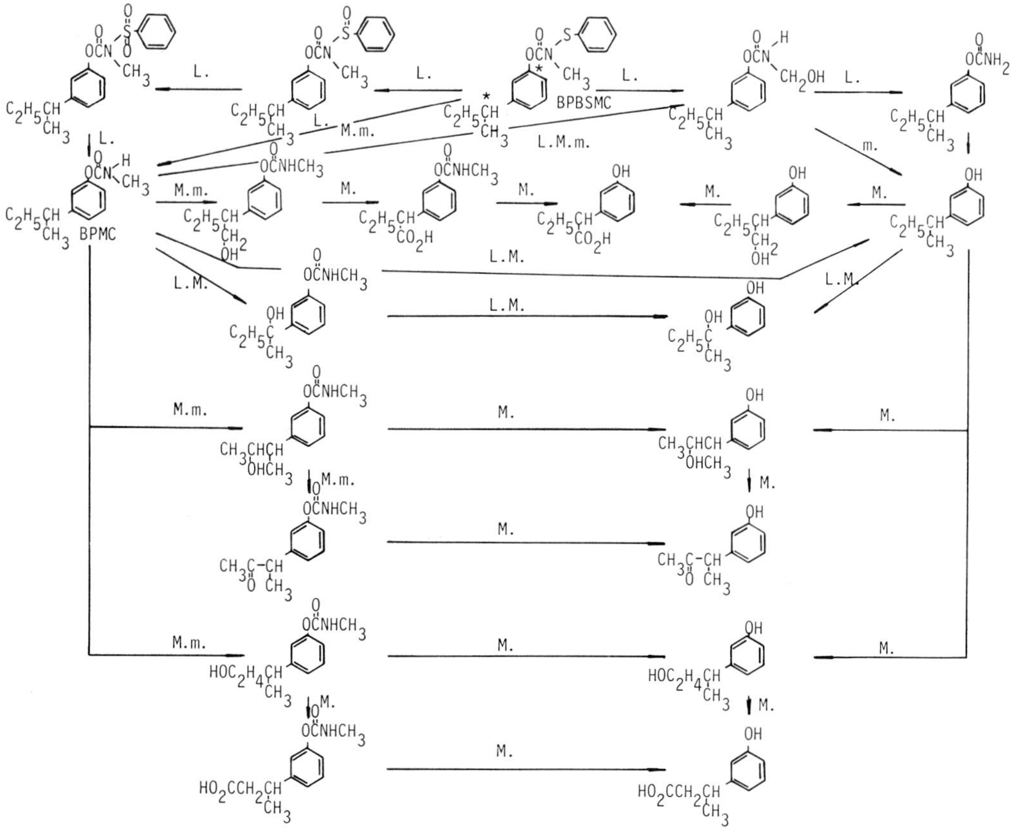

Light	Midday sunlight; UV lamp; sunlamp[154]
Mammal	Male albino rats[154]
Microsome	Liver microsome-NADPH system[154]

PROPOXUR (BAYGON)

When the *tert*-carbon of 2-isopropoxy group of propoxur is oxidized to form a hemiketal, mono-*N*-methylcarbamoylcatechol is produced as a hydrolyzed metabolite that can normally be detected as 2-isopropoxyphenol by hydrolysis of the carbamoyl ester linkage, although this phenol has still not been detected. The 5-position of the phenyl ring is selectively metabolized in insects and their microsomes.

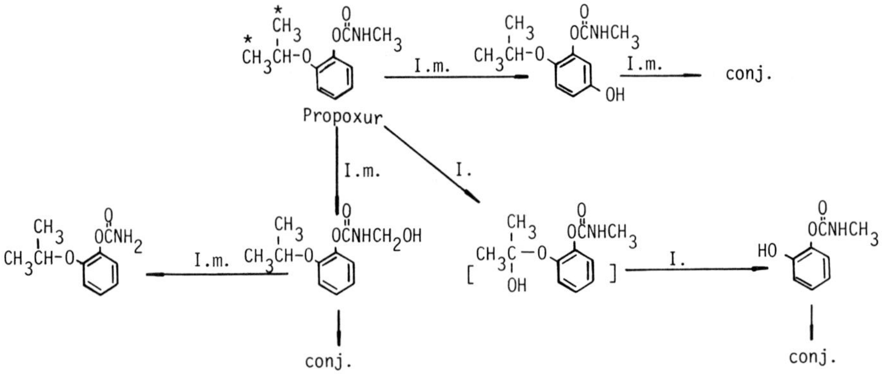

Insect	Housefly[155]
Microsome	Mosquito homogenate-NADPH system [*Culex pipiens fatigans* susceptible (S-Lab)][156]

MPMC (MEOBAL)

One of the methyl groups substituted at the 4-position of the phenyl ring in MPMC is sensitively oxidized to its corresponding carboxylic acid via benzyl alcohol in the major metabolic degradation. On the other hand, the methyl group at 3-position undergoes a one-step oxidation to the corresponding benzyl alcohol in mammals.

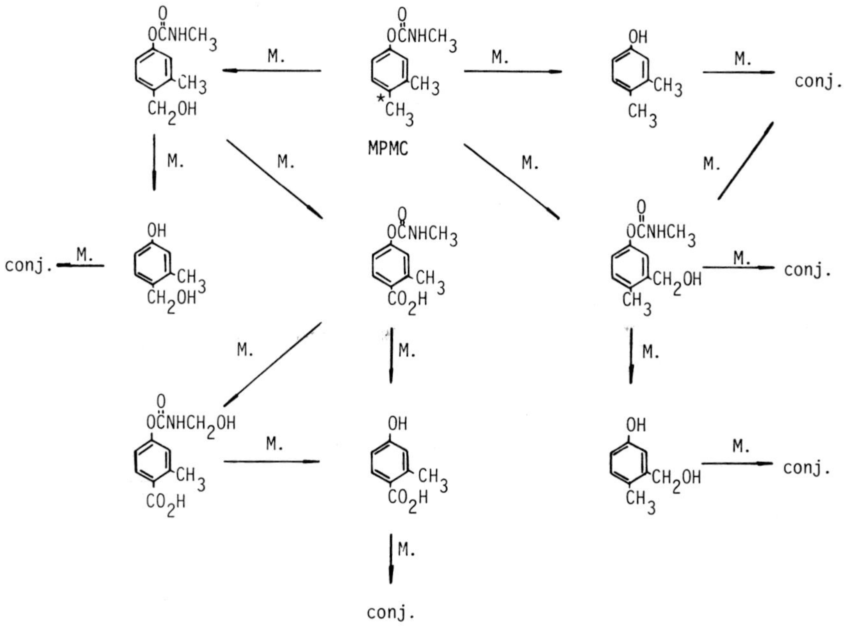

Mammal Rats; Mice[153]

UC-34096

Hydrolysis of the carbamoyl ester linkage and the deamination of the amidine moiety are more predominant than oxidation of the methyl group substituted in the phenyl ring in the metabolism of UC-34096. UC-34096 is easily decomposed in water.

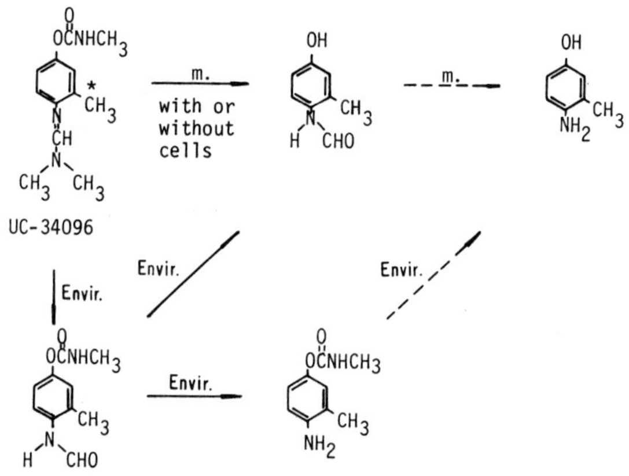

Microsome	Human embryonic lung cells culture (L-132 strain)[152]
Environment	Water[152]

MEXACARBATE (ZECTRAN)

Hydrolysis of the carbamoyl ester linkage, N-demethylation via *N*-formyl intermediates of the 4-substituted dimethylamino moiety, and hydroxylation of the *N*-methyl group in the carbamoyl moiety are observed. 3,5-Dimethyl-*p*-hydroquinone and 3,5-dimethyl-*p*-aminophenol are the characteristic metabolites in plants.

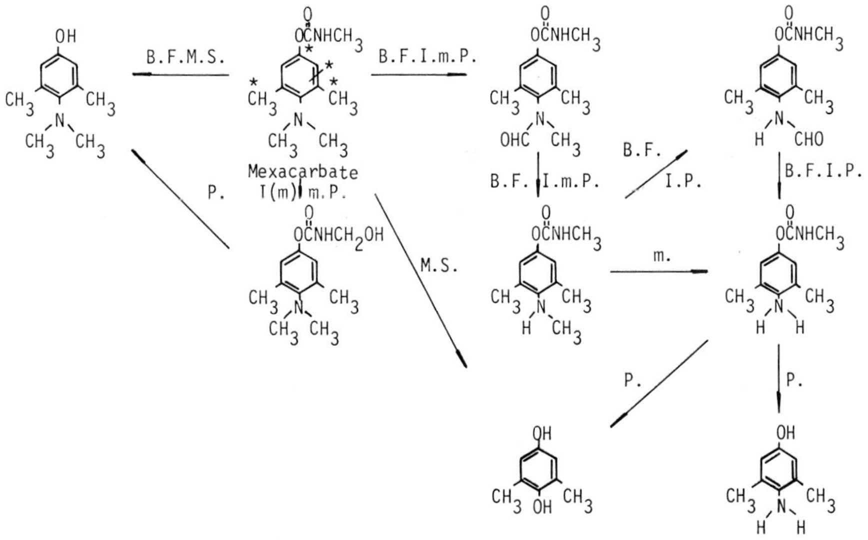

Bacteria	Thirty-five aerobic organisms (ex. HF-3 from housefly gut bacteria)[157]
Fungi	ex. *Trichoderma virdide* fungus[157]
Insect	$NADPH_2$-requiring enzyme system from housefly abdomen[158]
Mammal	[158]
Microsome	Rat liver microsomes[158]
Plant	Beans; Broccoli; Corn; Soybeans[158]

EP-475 AND PHENMEDIPHAM

EP-475 and phenmedipham are hydrolyzed to ethyl-*N*-(3-hydroxyphenyl) carbamate and methyl-*N*-(3-hydroxyphenyl) carbamate, respectively, both of which are subsequently cleaved to 3-aminophenol, and then N-acetylated. All of these metabolites undergo conjugation reactions.

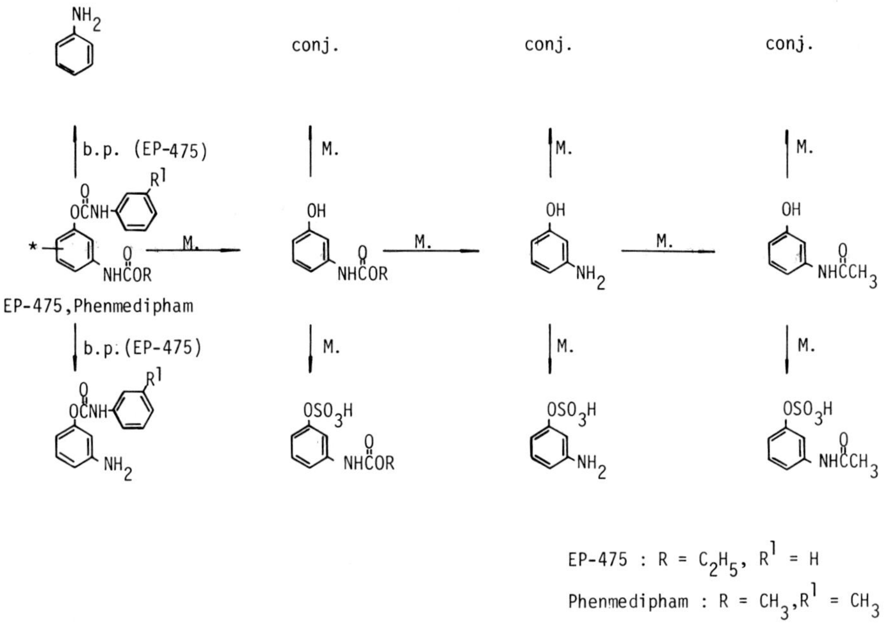

Mammal	Male white S-D rats[159]
Microsome	Rat liver microsomes[159]
Blood plasma	Human, rat, chicken, and cow[159]

CARBARYL (NAC, SEVIN)

Various metabolic reactions of carbaryl are shown for some metabolic systems.

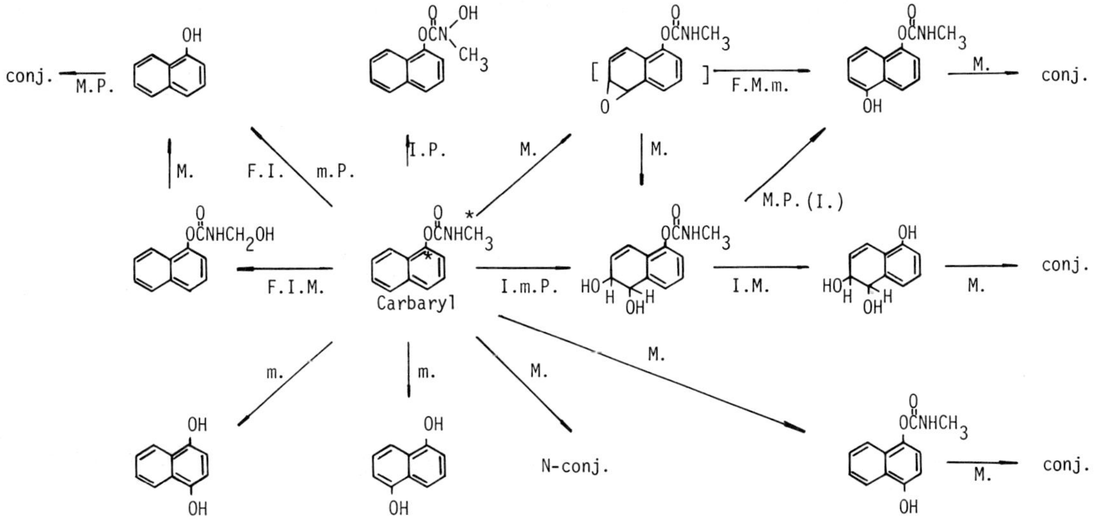

Mammal	Dog; Guinea pig; Man; Monkey; Pig; Sheep[153]
Plant	Tobacco cells in suspension culture (Xanthi X-D cell lines)[152]
Fungus	*Gliocladium roseum* (Link) Thom.[160] *Aspergillus flavus* Link. ex Fries; *A. fumigatus* Fresenius; *A. niger* van Theighem; *A. terreus* Thom.; *Aspergillus* sp.; *Fusarium oxysporum* Schlectendahl; *F. roseum* Link.; *Fusarium* sp.; *Geotrichum candidum* Link.; *Gliocladium roseum* (Link) Thom.; *Helminthosporium* sp.; *Mucor racemosus* Fresenius; *Mucor* sp.; *Penicillium roqueforti* Thom.; *Penicillium* sp.; *Rhizopus* sp.; *Trichoderma viride* Per. ex Fries[161]
Microsome	Human embryonic lung (HEL) cells[162]
Insect	Cabbage looper larvae: *Trichoplusia ni*[155] Alfalfa leaf cutting bee: *Megachile pacifica* (Panzer) (=*rotundata*)[163]

CARBOFURAN (FURADAN)

The main degradation reaction of carbofuran occurs at the 3-position of the dihydrobenzofuran structure to give hydroxyl and the resulting ketone derivatives, in addition to *N*-methyl hydroxylation and hydrolysis of the ester linkage. All the metabolites undergo conjugation reactions.

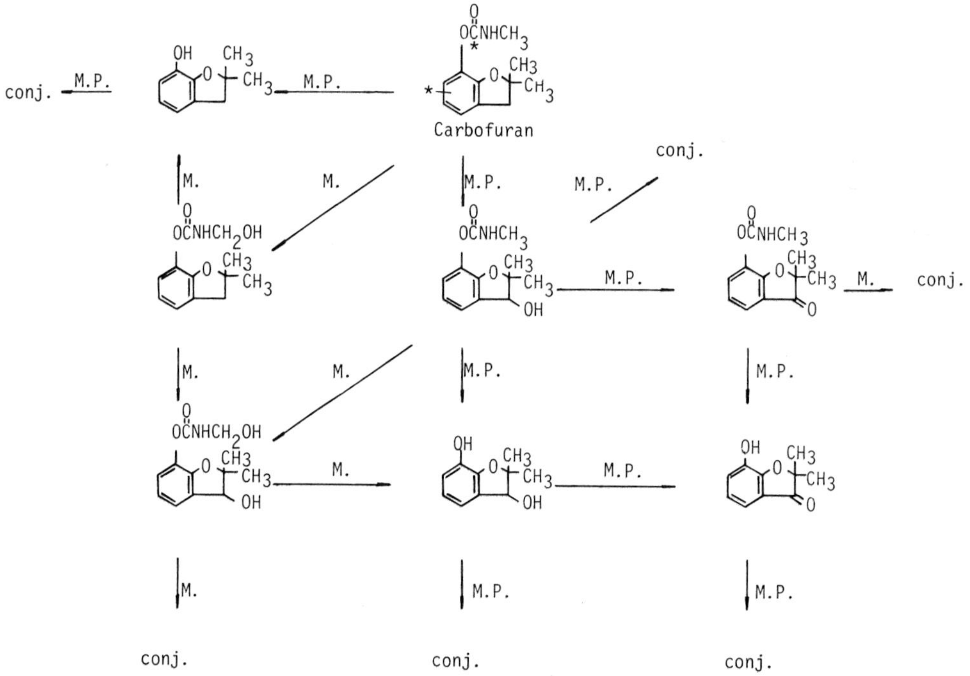

Mammal	Rat[153]
Plant	Alfalfa Narragansett variety; Bean[164] Tobacco[165] Mugho pine: *Pinus mugo* Turra[166] Bean[167]
Mammal	Rat[153] Female rats (Cox-SD variety)[167]

MOBAM

Mobam is easily hydrolyzed in mammals to its phenol, which then undergoes a conjugation reaction, although the benzothiophene ring undergoes no metabolic degradation reaction.

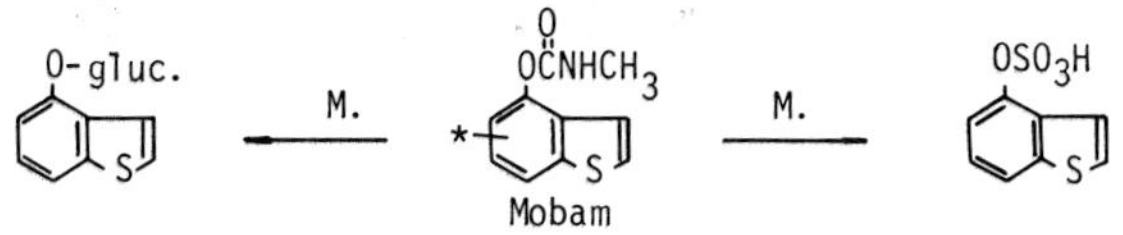

Mammal Rat[153]

14

Phenylureas and Related Compounds

SIDURON

Siduron is hydroxylated on both the phenyl and methyl cyclohexyl substituents. In the latter, hydroxylation may occur on either the cyclohexyl ring or methyl group giving secondary and primary alcohols, respectively, in plants.

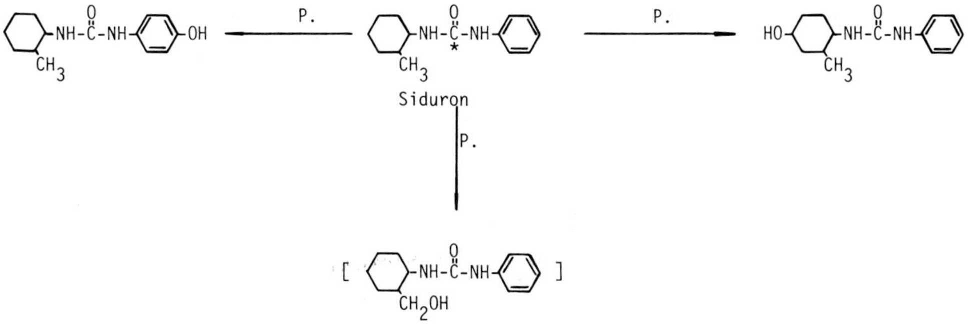

Plants	Merion Kentucky bluegrass: *Poa pratensis* L.[168]

MONURON

N-Hydroxymethylation of monuron causes the formation of a β-D-glucoside, and other polar, unknown methanol-soluble metabolites and insoluble residues in higher plants. The terminal demethylation product is *p*-chlorophenylurea. Chlorine and hydrogen atoms on the phenyl ring are replaced by a hydroxyl group; other rearranged products are also formed in light.

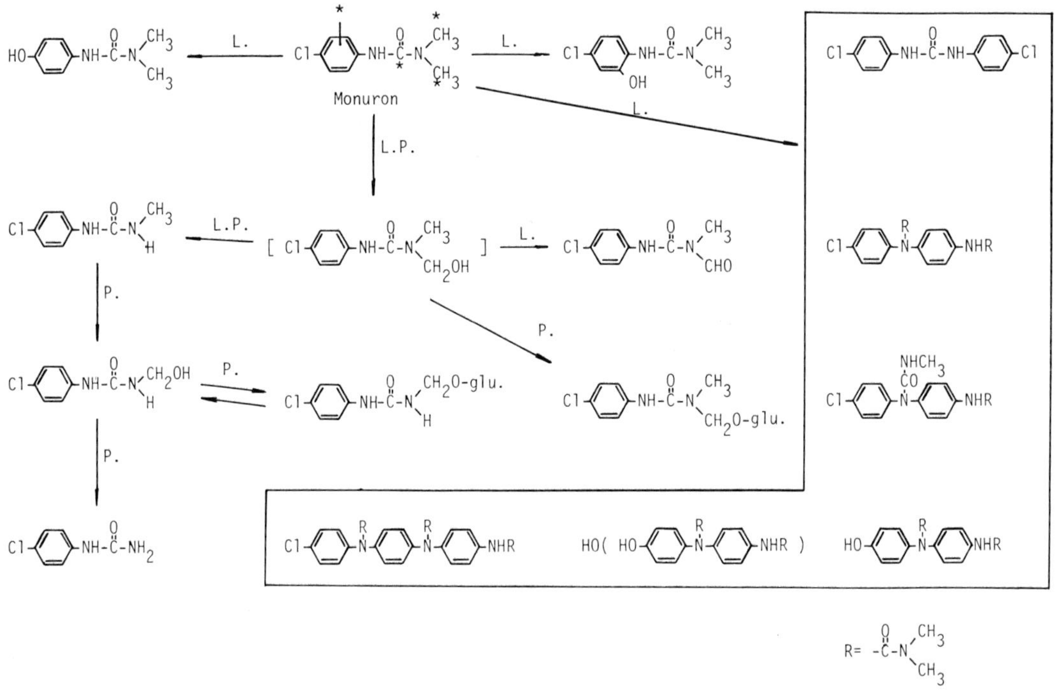

Plant	Fully expanded greenhouse-grown cotton: *Gossypium hirsutum* L.[169]
Light	Photolysis by Rayonet photoreactor[170]

MONOLINURON

The methyl group in monuron is replaced by a methoxy group in monolinuron. This methoxy group is resistant for degradation, so the metabolites are *N*-hydroxymethyl, its conjugate, and demethylated products in bacteria.

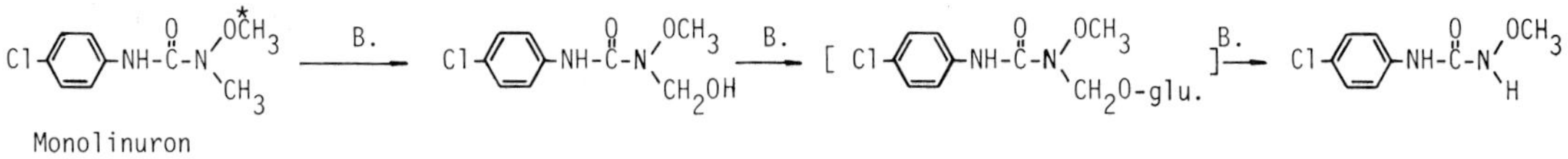

Bacteria *Chlorella pyrenoidosa*[171]

BUTURON

Because of the *N*-*sec*-butynyl group, normal demethylation is not observed, but reduction, hydroxylation, and alcoholysis take place at the triple bond, and hydroxylation of the ureido moiety gives the corresponding methylurethane; this is followed by methylation at the nitrogen atom in soil. In water, *N*-formyl-*p*-chloroaniline and dechlorinated hydroxyphenylurea are detected in the degradation products.

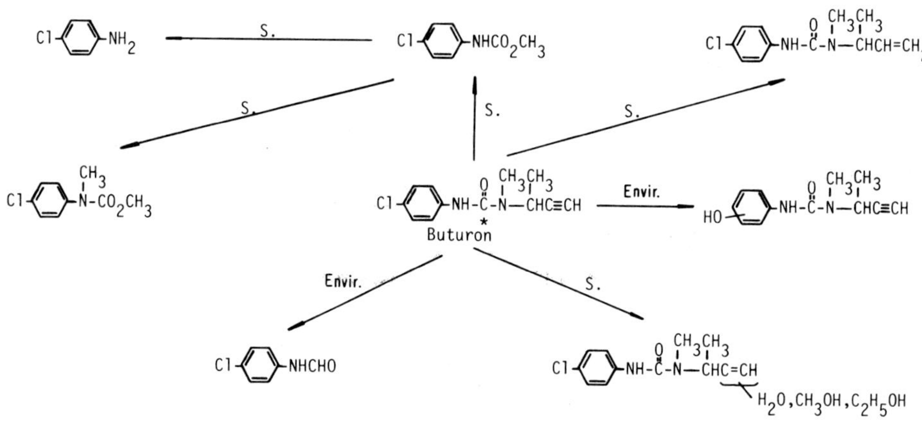

Soil	Soil (pH 7.0)[172]
Environment	Leaching water[172]

CLEARCIDE

N-(3-Chloro-4-chlorodifluoromethylthio)phenylurea and 3-chloro-4-chlorodifluoromethylthioaniline are produced by hydrolysis of monodemethyl clearcide which is the first degradation product. The sulfur atom of the chlorodifluoromethylthio group of the mother compound and its metabolites, which have *N*-methyl groups in the ureido moiety, are oxidized to the corresponding sulfoxides and sulfones in soil.

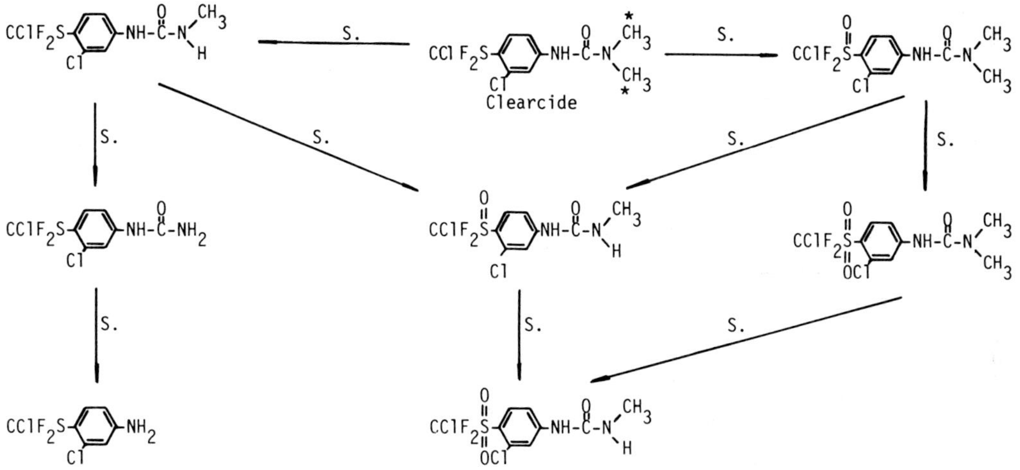

Soil Paddy field soil: Alluvial clay loam (pH 5.4); Alluvial loam (pH 6.0); Volcanic ash silty loam (pH 5.9)[173]

PH-6040 (OMS-1804; DIFLUBENZURON)

Degradation of PH-6040 is apparently initiated by cleavage at both the N-1 and C-2 bonds to give 2,6-difluorobenzamide and 4-chloroformanilide, and at the N-1 and C-1 bonds to give 2,6-difluorobenzoic acid and 4-chlorophenylurea. 2,6-Difluorobenzamide is ultimately hydrolyzed to 2,6-difluorobenzoic acid and conjugated into predominantly water-soluble products. Both 4-chloroformanilide and 4-chlorophenylurea are further decomposed to 4-chloroaniline, which is both acetylated and methylated biologically.

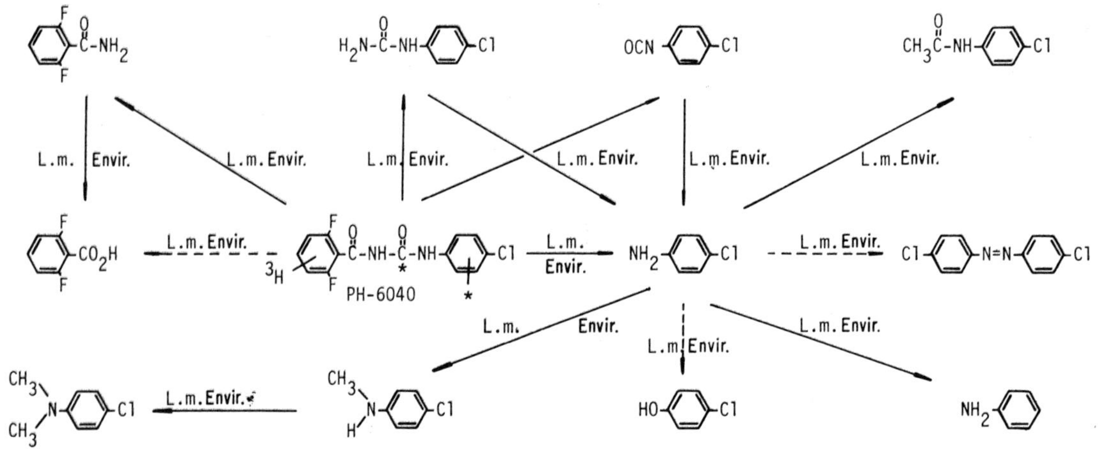

Light	Photodegradation in methanol and aqueous dioxane[174]
Microsome	Sheep liver microsomes[174]
Environment	A model ecosystem = H_2O; *Sorghum vulgare;* Forth instar salt marsh caterpillars: *Estigmene acrea;* Plankton: *Daphnia magna;* Alga: *Oedogonium cardiacum;* Snail: *Physa* sp.; Mosquito larvae: *Culex pipiens quinquefasciatus;* Fish: *Gambusia affinis;* Soils[174]

BATH (BENZOIC ACID 2-[2,4,6-TRICHLOROPHENYL]HYDRAZIDE)

Three major degradation reactions occur in photolysis: (1) Reductive loss of chlorine atom, (2) loss of either one or two nitrogen atoms, and (3) benzoylation.

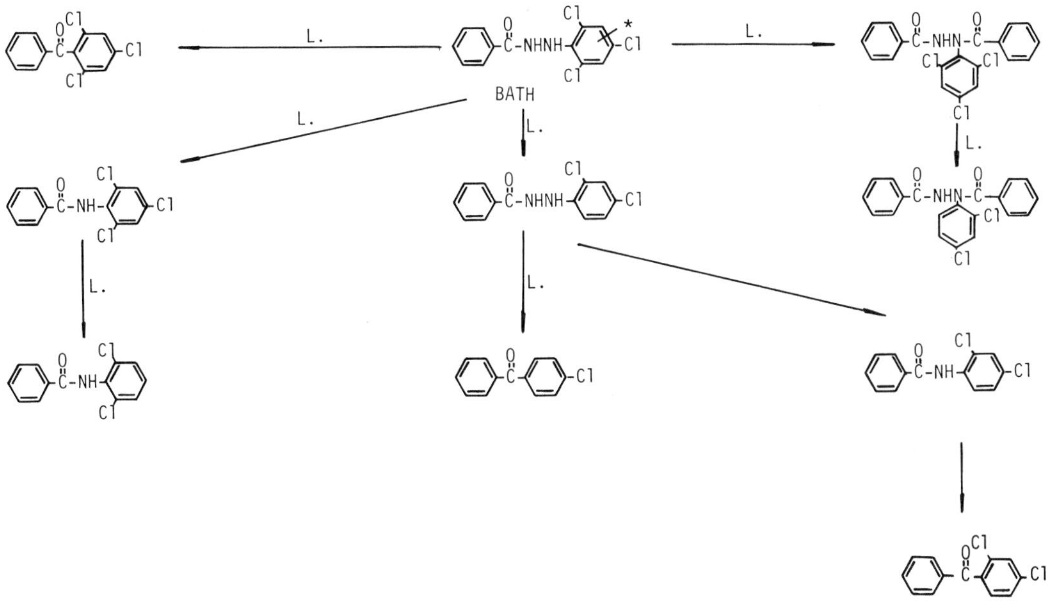

Light UV lamp: Rayonet RPR model on glass beads[175]

THIADIAZURON

Hydroxylation at the 4-position of the aniline moiety is the important metabolic pathway. The 1,2,3-thiadiazole ring is stable, but phenylurea which cannot be formed by dealkylation is detected as a metabolite in mammals. This metabolite represents a novel cleavage for substituted urea herbicides.

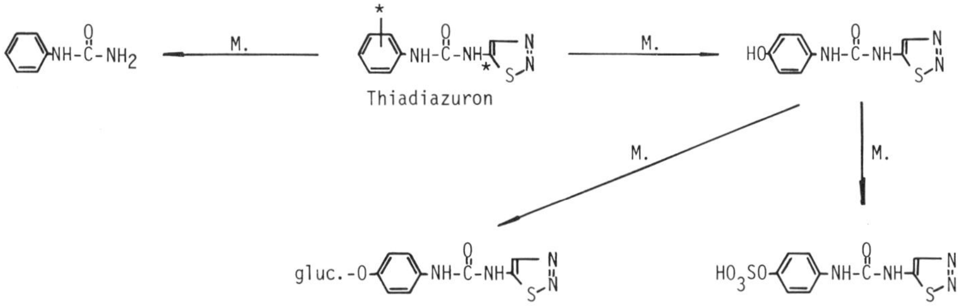

Mammal Female and male albino S-D rats[176]

15

Phosphonothiolates and Phosphonothioates

LEPTOPHOS

A main metabolic pathway is hydrolysis of the P–O bond which results in the corresponding phenol and phosphonothioic acid; the latter is oxidized to give phosphonic acid in mammals.

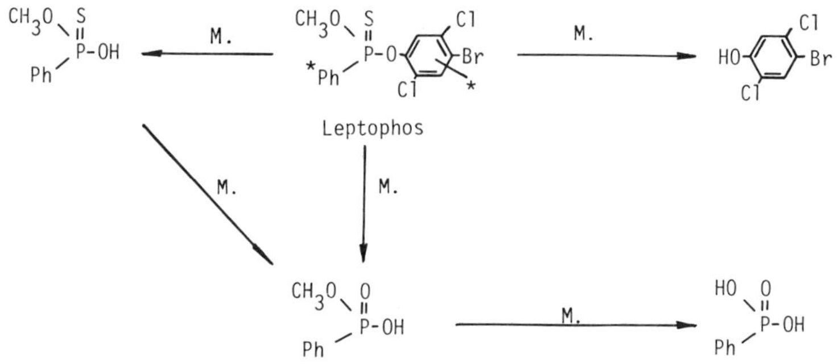

Mammal Adult Holtzman rats[177]

INEZIN

In addition to hydrolyzed products, a hydroxylation product at the meta position of the phenyl ring is detected especially in fungi.

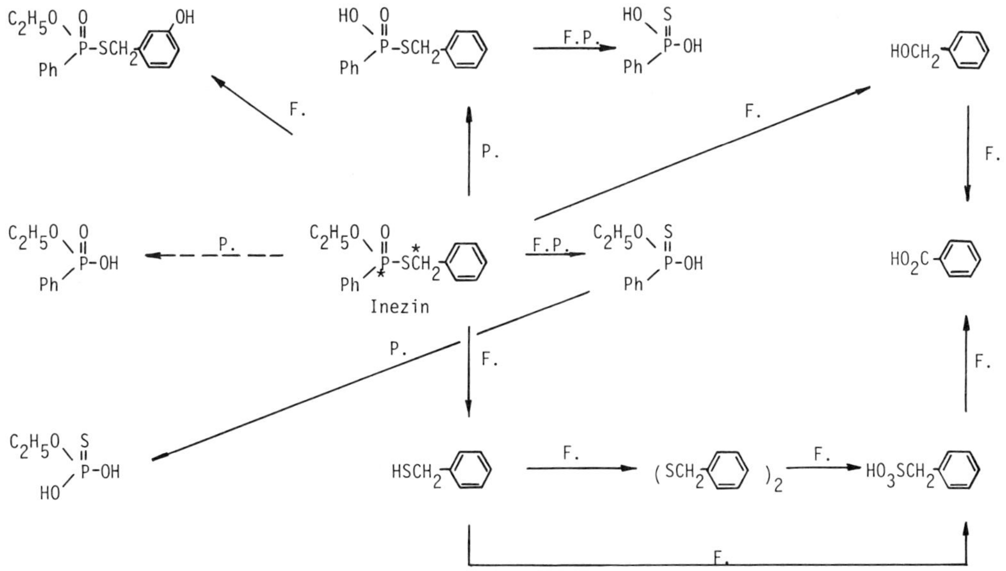

Plant Rice[178]

Fungi *Pyricularia oryzae*[179]

DYFONATE (FONOFOS)

Formation of oxon and hydrolyzed products of dyfonate are observed. It is of interest that methyl esters of phosphonic acid and hydroxylated phenyl methyl sulfones are detected, but it is not known whether two methyl phosphonate esters are represented as *in vivo* forms or do they arise from nonmetabolic products. The oxon occurs in only very small amounts, so cleavage of dyfonate to ETP and subsequent transformation to EOP are probably predominant for EOP formation. Toxicity of dyfonate and its degradative metabolites to rats, houseflies, and *Gambusia affinis,* and I_{50} of bovine erythrocyte AchE are listed (LD_{50}, LC_{50},M) in the table. Only dyfonate and its oxon, but none of the other identified metabolites, show any appreciable toxicity to several test animals.[180]

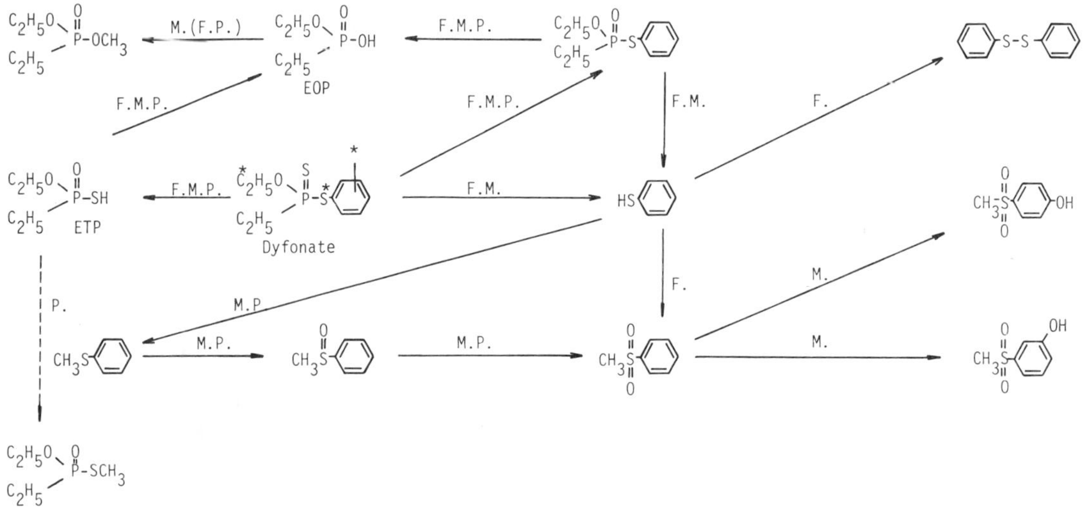

Plant	Irish potato[180]
Mammal	Albino rats[181]
Microsomes	Liver microsomes[181]
Fungi	*Aspergillus flavus* Link.; *Aspergillus fumigatus* Fresinius; *Aspergillus niger* var. Tieghem; *Fusarium oxysporum* Schlechtendal; *Mucor altenans* var.

Tieghem; *Mucor plumbeus* Bonorden; *Penicillium notatum* Westling; *Rhizopus arrhizus* Fisher; *Trichoderma viride* Pers ex. Fries[182]

Soil fungus: *Rhizopus japonicus*[183]

N-2596

N-2596 has the same structure as dyfonate except that the phenyl group is substituted by a chlorine atom at the para position. N-2596 is metabolized by two different biotransformation pathways. One involves oxidation of *p*-chlorothiophenol to *p*-chlorophenylsulfonic acid, probably through the intermediate, sulfinic acid. Formation of the *S*-glucuronide of *p*-chlorophenylmercaptan provides evidence for yet another metabolic pathway in mammals. Toxicity (rat oral LD_{50}) and anticholinesterase inhibition (bovine erythrocytes AChE, I_{50}, M) for N-2596, its metabolites, and related compounds are listed in the table.[184] With the exception of the oxygen analog of N-2596, these compounds show less acute toxicity and are less potent inhibitors of bovine erythrocyte acetylcholinesterase than the parent insecticide.[184]

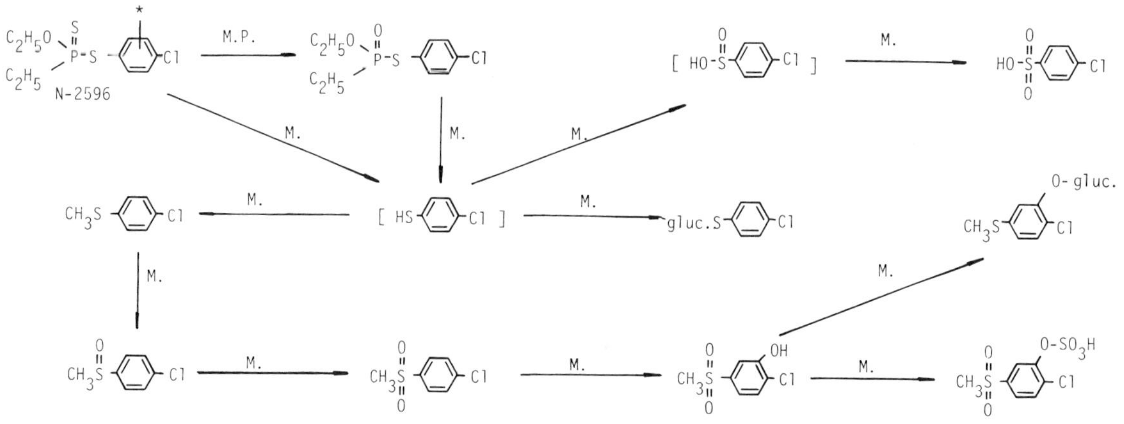

Mammal Female and male Simonsen albino rats[184]

Plant Broccoli; Brussels sprouts; Cabbage; Cauliflower (cole crops); Corn; Sugarbeets[185]

16

Phosphonates

TRICHLORFON

A typical metabolic pathway determines the toxic phosphate dichlorvos or the rearranged product of trichlorfon in insects, mammals, and plants. Of the compounds detected, only trichlorfon and dichlorvos are biologically active. Metabolism of dichlorvos can be seen in the metabolic maps of phosphate insecticides.

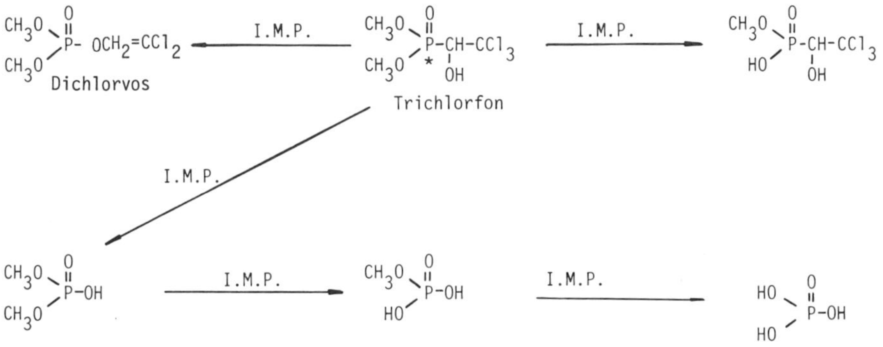

Insect	Adult lygus bugs: *Lygus hesperus* Knight; Tobacco bud-worm: *Heliothis virescens* F.; Green lacewing larvae: *Chrysopa carnea* Stephens[186]
Mammal	Female white rats[186]
Plant	Cotton leaves[186]

SURECIDE

In addition to hydrolysis of this insecticide, direct carboxylation occurs in the phenyl rings of the resulting *p*-cyanophenol during photolysis during which toxic hydrogen cyanide is liberated.

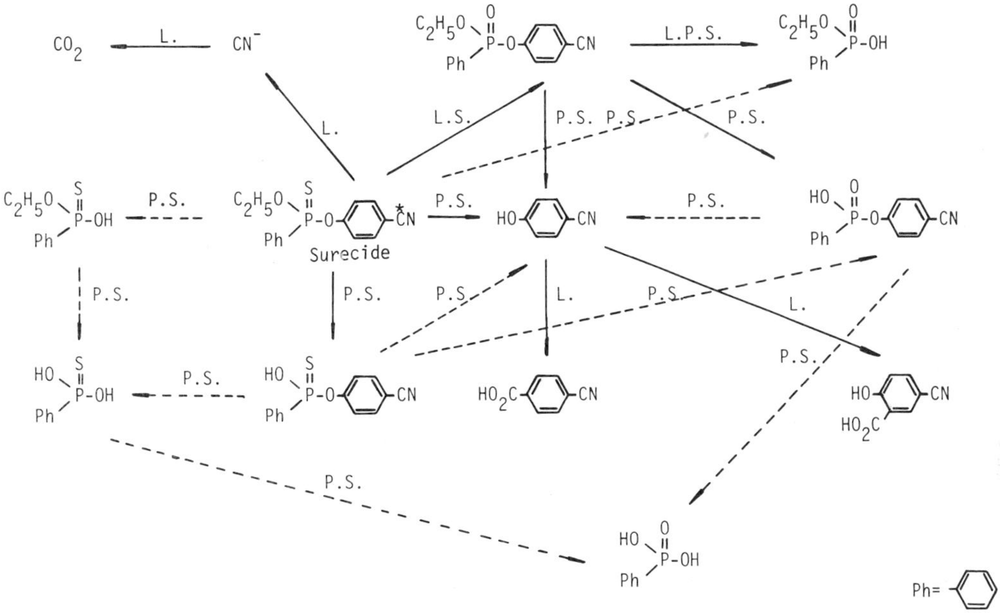

Plant	Bean: *Phaseolus vulgaris* L. 'Nagauzura'[187]
Soil	Loam soil (pH 6.5): Setagaya top soil[187]
Light	[187]

GLYPHOSATE (ROUNDUP)

The parent compound is N-methylated and degraded into three compounds at the initial metabolic sequence. These are further metabolized to N-methylated glycines and some phosphonic acids in plants, soil, and water.

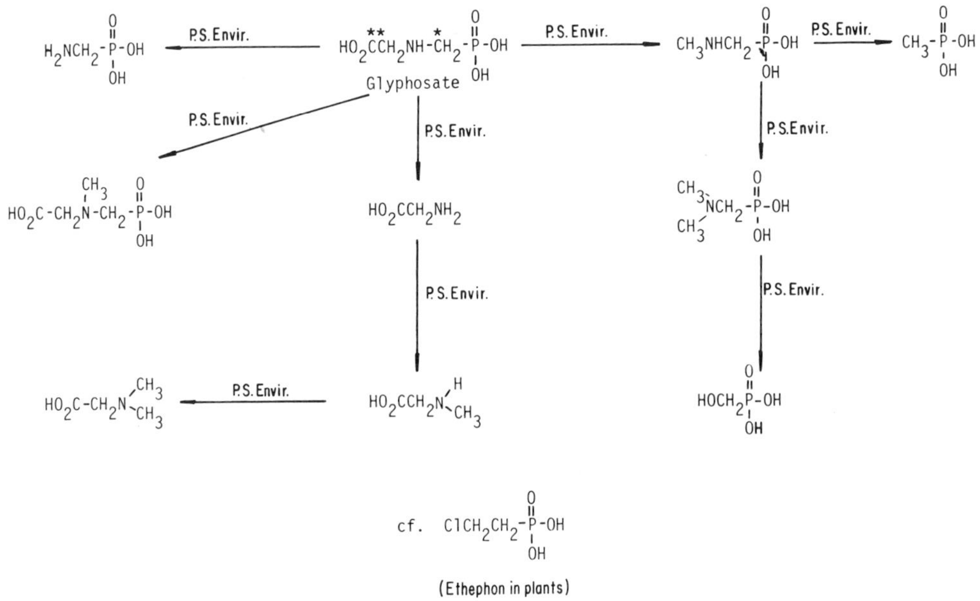

Plant	Montmorency cherry leaves[188]
Soil	Drummer silty clay loam; Norfolk sandy loam; Ray silt loam; Litonia sandy loam[189]
Environment	Soil water[189]

17

Phosphorothioamides

CREMART

In mammals the nitro group is reduced to amine and the methyl group is oxidized to carboxylic acid which has the basic structure of cremart, but 5-methyl-2-nitrophenol, formed by hydrolysis, might be reduced to aminophenol which is not identifiable in the environment. The methyl group of the nitrophenol is oxidatively metabolized to alcohol and carboxylic acid in mammals and plants.

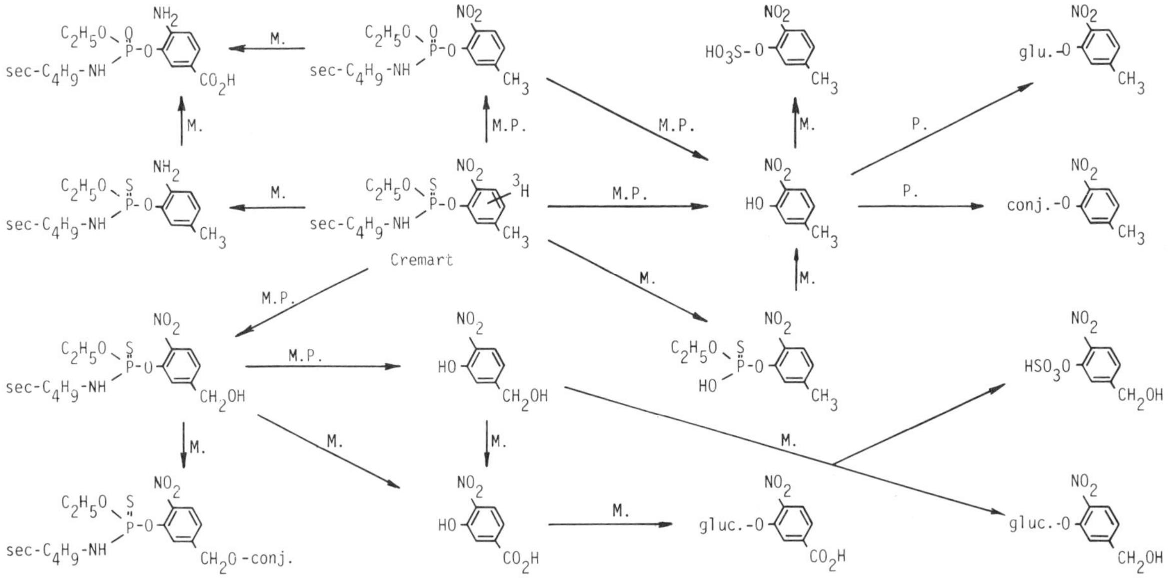

Mammal Female and male S-D rats[190]

Plant Bean: *Phaseolus vulgaris* L.; Rice: *Oryza sativa* L.; Carrot: *Daucus carota* L., var. *Stativa* DC[190]

S-2517

S-2517 has an N-C_3H_7-*i* group substituted in place of the *sec*-C_4H_9 group in cremart. No reductive degradation of the nitro group is observed, but oxidative degradation to oxon and to primary alcohols and aldehydes at the methyl group of the phenyl ring occurs during photolysis and in microsomal systems.

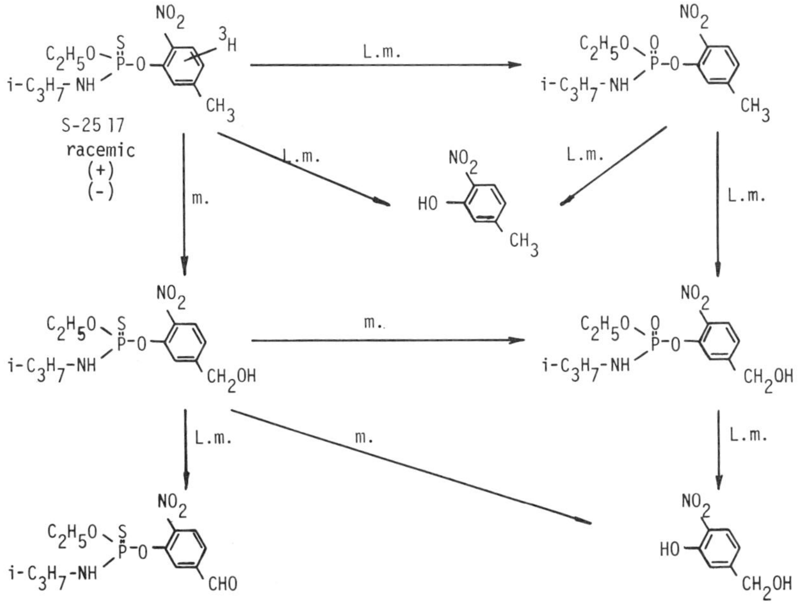

Microsome	Rat liver microsomes + NADPH system[191]
	Liver microsome function (Male rabbit: Albino Native Japanese strain)[192]
Light	UV light[192]

18

Phosphoramides, Phosphoramidothiolates, and Phosphorimides

NEMACUR (BAY-68138)

Residues of nemacur consist mostly of sulfoxides, sulfones, and sulfides in plants and microsomes.

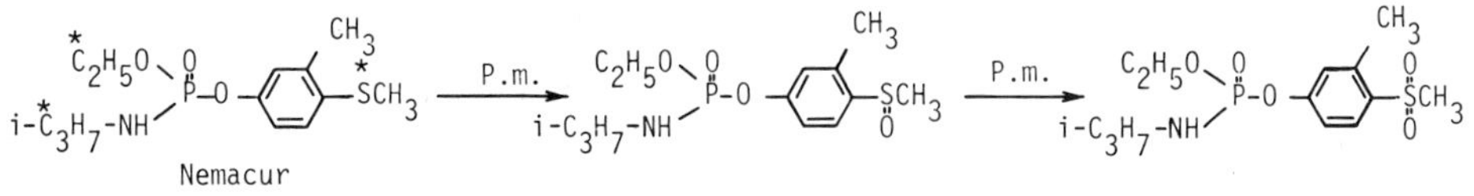

Plant	Citrus peel and pulp; Pineapple fruit; Cured tobacco; Peanut hulls; Peanut vines; Pineapple bran and forage; Peanut meat[193]
	Top crop bush beans: *Phaseolus vulgaris;* Rutgers tomatoes: *Lycopersicon esculentum;* Mammoth Jumbo peanuts: *Arachis hypogaea;* Kennebec white potatoes: *Solanum tuberosum*[194]
Microsome	Animal tissues (other than fat)[193]

CRUFOMATE

N-Demethylation, deamination, and hydroxylation at the methyl group bound to quartery carbon atoms are the first degradation step of crufomate, and further metabolism gives various oxidation products, including glucuronide conjugates in mammals.

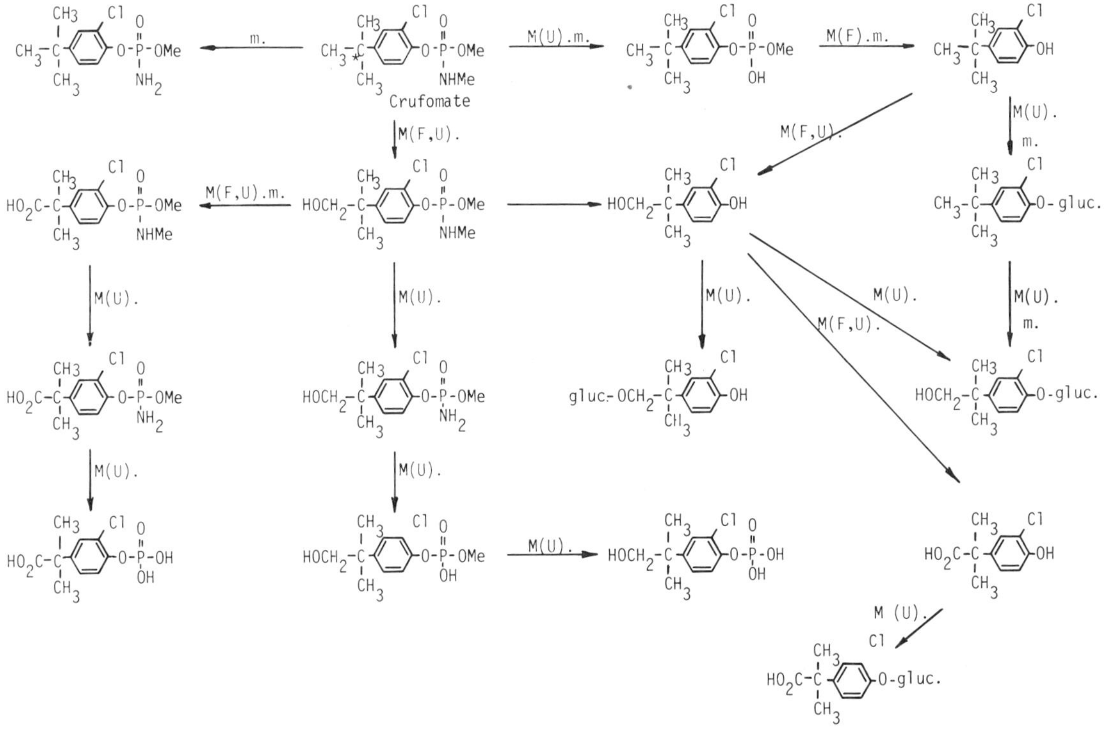

Mammal Male rats[195]

Microsome Intestine of rats (Incubation)[196]

METHAMIDOPHOS

The active intermediate derived from methamidophos is not sufficiently stable to allow its isolation. The structure of the active intermediate is a sulfoxide, i.e., *O,S*-dimethylphosphoramidothiolate S-oxide. However, such activation cannot be observed in phosphorothiolate triesters. It appears that the sulfoxide formation may not only be the activation mechanism for phosphoramidothiolate esters, but also the possible degradation mechanism for phosphorothiolate esters, because the ester oxygen atom is a weaker electron donor in nature than the nitrogen atom. The major difference in the metabolism of I and II in mice appears to be in larger amounts of methamidophos formed from II compared to I. Methamidophos is highly toxic to mice (LD_{50} 14 mg/kg) and probably is the agent responsible for intoxication when the mouse is treated with either I or II.

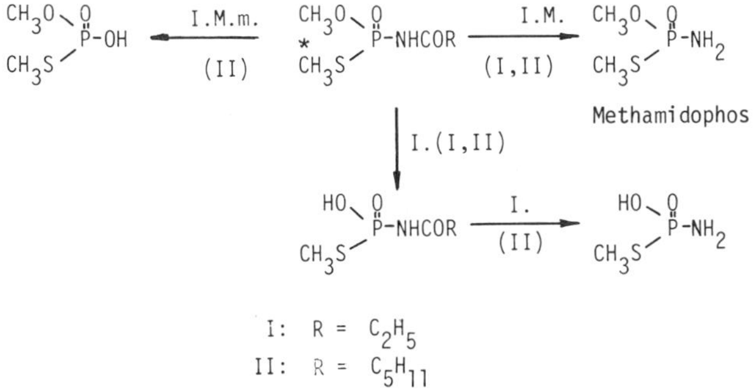

Insect	Housefly, 3-day-old female susceptible: SNAIDM strain[197]
Mammal	Female Swiss white mice[197]
Microsome	Mouse liver microsomes[197]

STAUFFER R-16661

Although R-16661 is usually quite stable in the cotton plant and the housefly, it is slowly metabolized in these biological systems to give a 4-keto derivative which has relatively poor insecticidal activity and high mammalian toxicity.

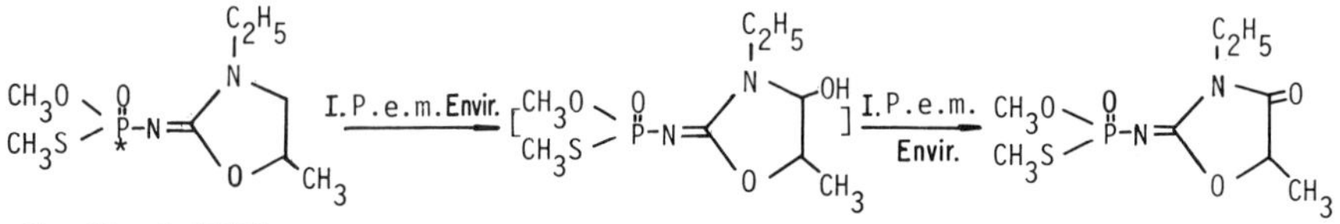

Stauffer R-16661

Insect	Housefly: *Musca domestica* L.
Enzyme	Housefly (*Musca domestica* L.[198])
Microsome	Mosquito larval and mouse liver enzyme preparations[198]
Plant	Cotton: Deltapine Smooth[198]
Environment	Udenfriend model system[198]

CYCLOPHOSPHAMIDE

Replacement of the chlorine atom by a hydroxyl group, N-dealkylation, and oxidation at the alpha position to nitrogen in the six-membered heterocyclic ring take place in addition to cleavage of the C–N bond of the ring by hydrolysis to give its corresponding alcohol and carboxylic acid in mammals.

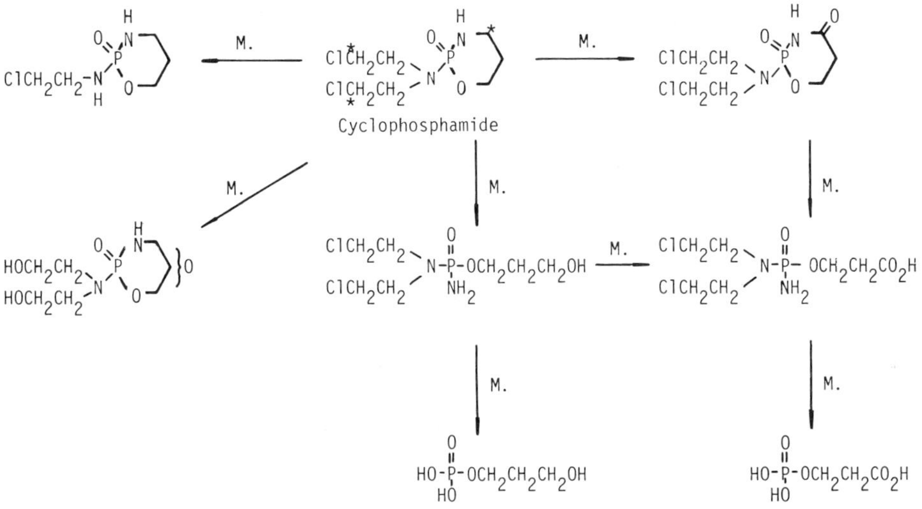

Mammal Wether sheep urine and feces[199]

CYOLANE (PHOSPHOLANE)

The only urinary and tissue metabolite that occurs in significant amounts has been identified as the thiocyanate ion. This shows that cyolane is initially hydrolyzed to 2-imino-1,3-dithiolane which is degraded to form the thiocyanate ion in mammals.

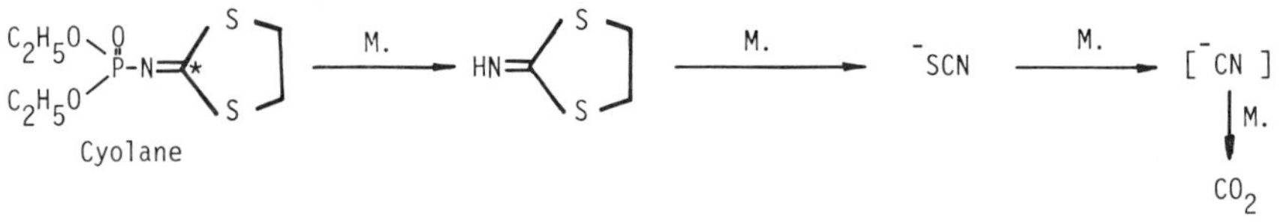

Mammal Royal Hart Wistar male rats[200]

MEPHOSFOLAN

Mephosfolan is analogous to cyolane, possessing a methyl group at the 4-position of the 2-imino-1,3-dithiolane ring, and undergoes oxidation at the methyl group to form primary alcohol in addition to the formation of the thiocyanate ion in plants.

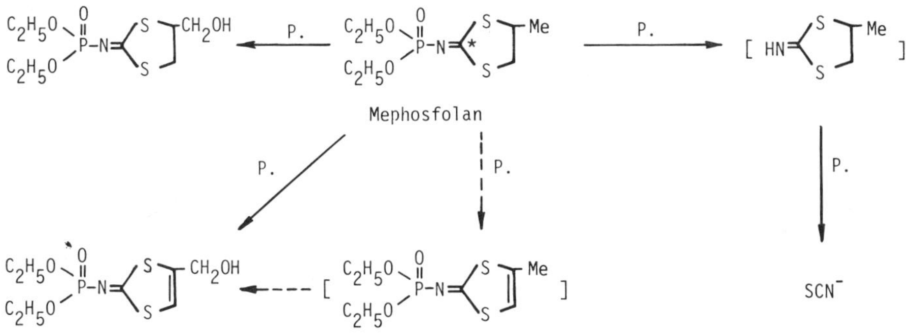

Plant Cotton: Deltapine smooth leaf variety[201]

19

Phosphates

GC-6506

Oxidation of GC-6506 to sulfoxide in plants is extremely rapid. Some further oxidation to sulfone occurs but at a greatly reduced rate. Initial decomposition of the toxic compounds involves cleavage at both *O*-Me and *P*-*O*-phenyl linkages. The failure of any nontoxic metabolite to accumulate and the rapid increase in unextractable radioactivity indicates that the decomposition products also are somewhat unstable and readily degraded to simple fragments, including inorganic phosphates in plants. Sulfoxide and sulfone are toxic to the adult boll weevil.

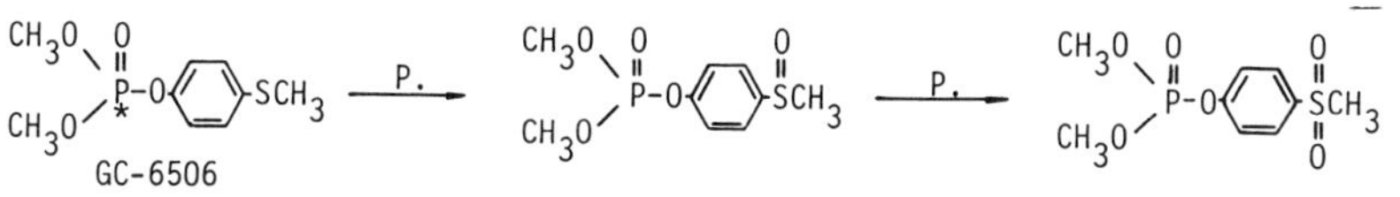

Plant Cotton: Deltapine smoothleaf variety[202]

DICHLORVOS

Dichlorvos, one of the active metabolites of trichlorfon already mentioned, is hydrolyzed to give dimethyl phosphate and dichloroacetaldehyde. The latter is subsequently reduced to β,β-dichloroethyl alcohol, characteristically converted by rats when administered intraperitoneally. After hydrolysis to a two-carbon fragment, dichloroacetaldehyde is able to enter into a pathway of intermediary metabolism, and CO_2 is the major radioactive metabolite.

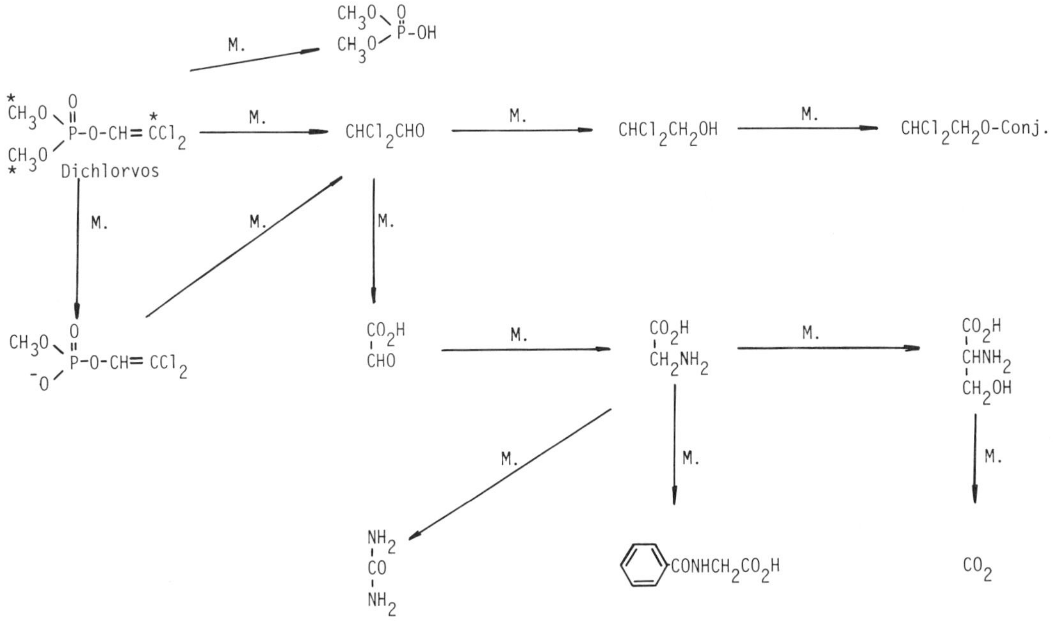

Mammal Adult male rats[203]

Adult Carworth Farm E strain rats; CF1 strain mice[204]

DIMETHYL AND TETRACHLORVINPHOS

Dimethyl and tetrachlorvinphos undergo hydrolysis and dealkylation resulting in dimethyl hydrogen phosphate and α-chloroacetophenones; the latter are further decomposed by oxidative demethylation and glutathione-dependent demethylation to their corresponding benzoic acids, and by another route of glutathione complex formation yield the corresponding mandelic acids in mammal and microsomal systems.

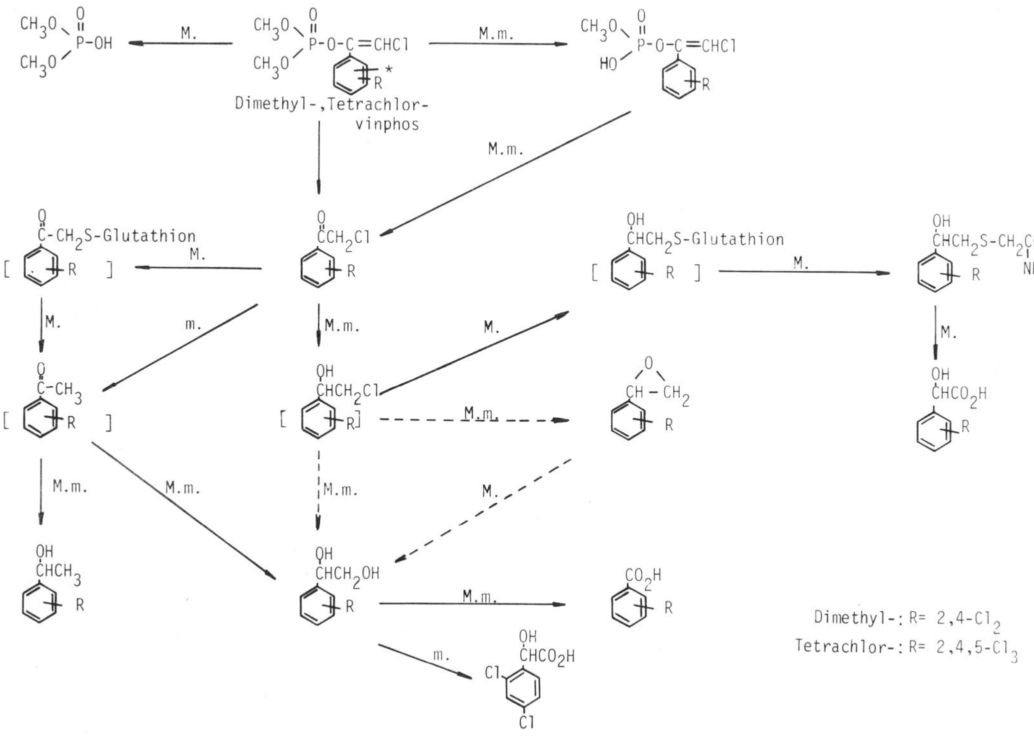

Mammal Young adult CFE rats; Male Beagle dogs[205]

Microsome Liver microsomes of rats and dogs[205]
Single comb White Leghorn hen liver enzyme preparation[206]

PHOSPHAMIDON

Hydrolysis of the *P-O*-Me bond, leading to the nontoxic polar metabolite, *O*-demethylphosphamidon, is a minor pathway of degradation of phosphamidon. Oxidative N-dealkylation is the only pathway operative in plants and mammals that leads to the toxic metabolites of phosphamidon. Displacement of chlorine by hydroxyl is one pattern of metabolism by hepatic microsomal enzymes that is also observed in mammals.

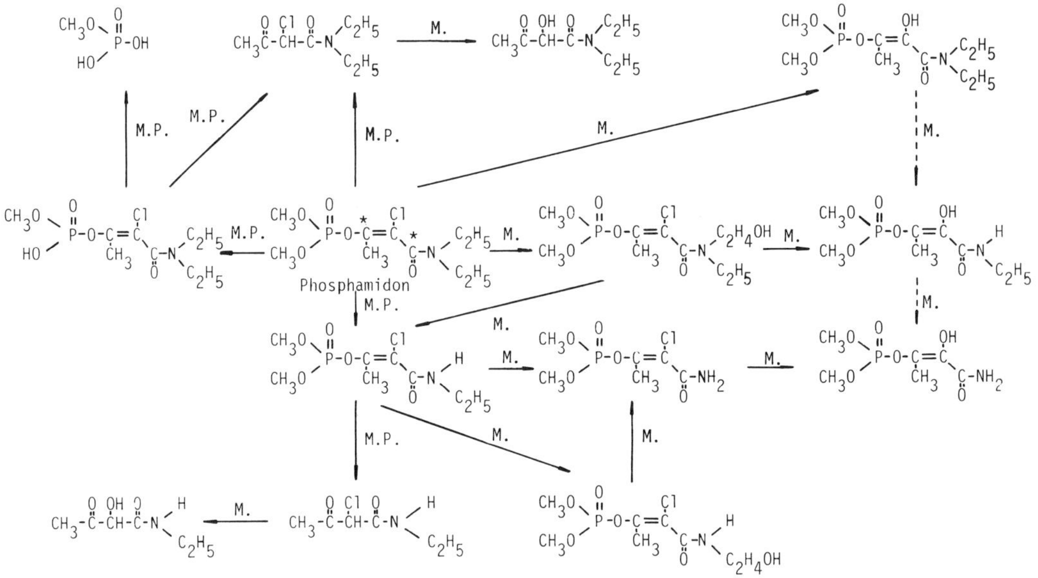

Mammal	Rats; Lactating cows; White mice; White rats; Goats[207]
	Female and male S-D rats[208]
Plant	French bean seedling: *Phaseolus vulgaris;* Cotton[207]
Microsome	Rabbit and rat microsomes[208]

20

Phosphorothiolates

KITAZIN P

A main pathway of kitazin P in plants is hydrolysis resulting in diisopropylhydrogen phosphorothiolate, isopropyldihydrogen phosphate, and phosphoric acid. Isomerization of kitazin P to phosphorothioate in plants and hydroxylation of the phenyl ring in fungi have been observed.

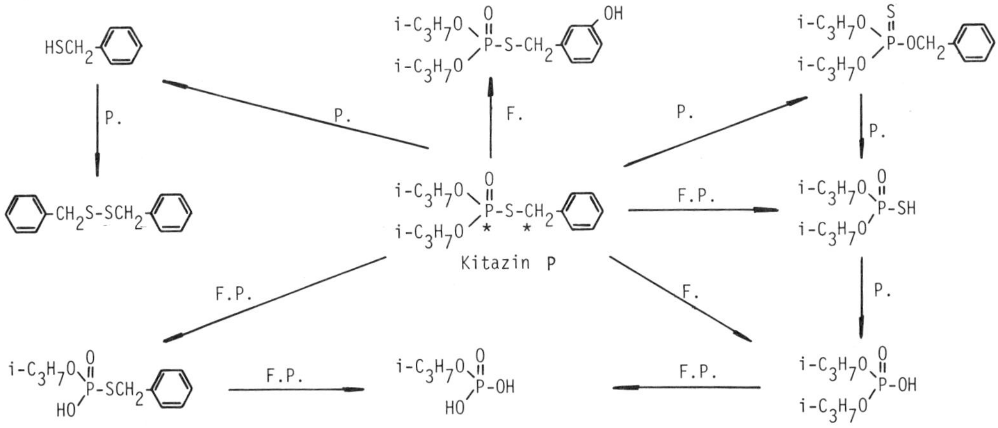

Fungi	*Pyricularia oryzae*[209]
Plant	Rice: 'Asahi' variety[210]

R-3828

As observed in various organophosphorus pesticides, R-3828 is degraded by hydrolysis, and the resulting benzhydrol is oxidized to benzophenone. One of the phenyl groups which does not possess a chlorine atom at the para position in both alcohols and ketones is selectively hydroxylated at this position in mammals.

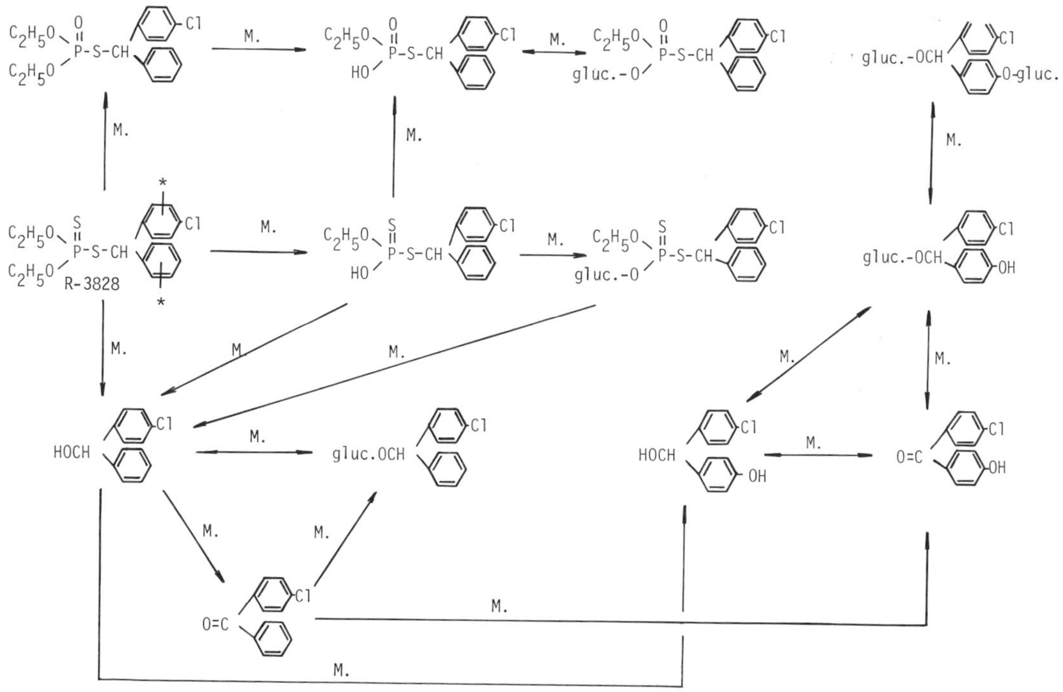

Mammal Steer[211]

21

Phosphorothioates

PARATHION

Although the main degradation reaction of parathion is preceded by oxidation to oxon and by hydrolysis to *p*-nitrophenol, reduction to aminoparathion and oxidation to aminoparaoxon have been observed in microsomes. Parathion is also reduced to aminoparathion through nitrosoparathion and hydroxyamino-parathion successively in plants.

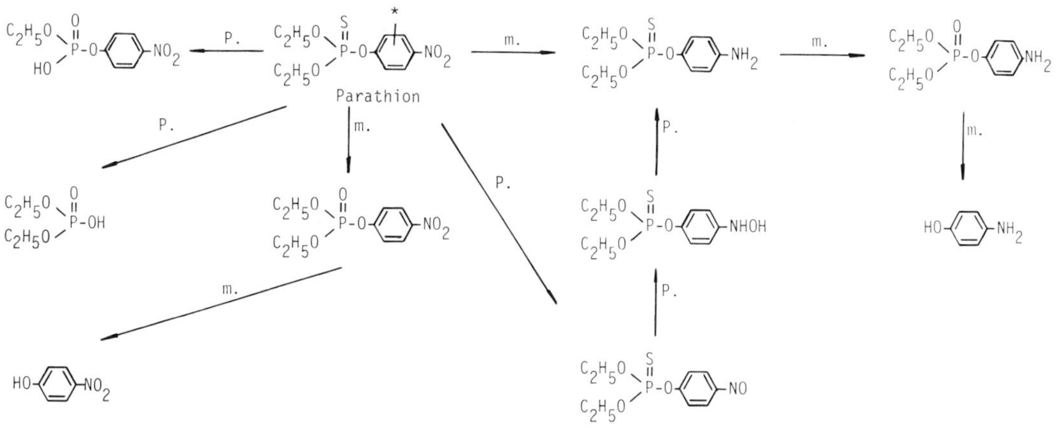

Microsomes	Female albino S-D rat liver cells; Rat liver mitochondrial subfraction[212]
Plant	Spinach homogenate[213]

SUMITHION (FENITROTHION)

Under most degradation conditions in the environment, fenitrothion yields degradation products; their toxicities can also be assessed. Almost all of the possible structures of the metabolites are determined in the case of fenitrothion.

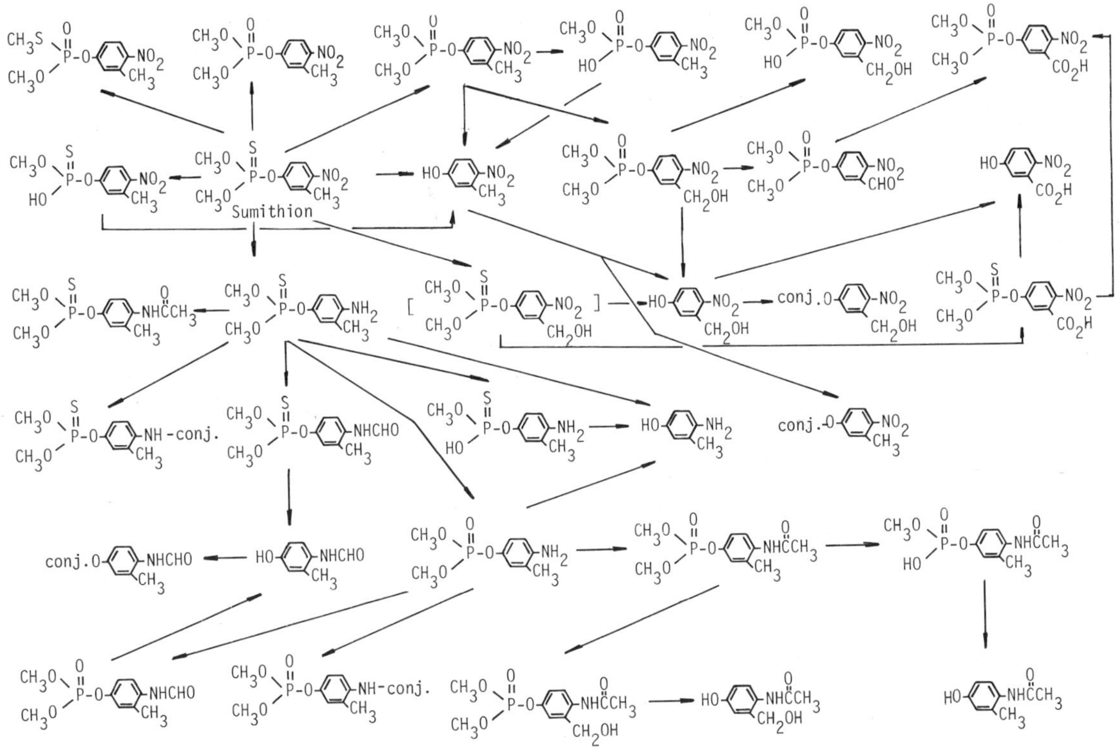

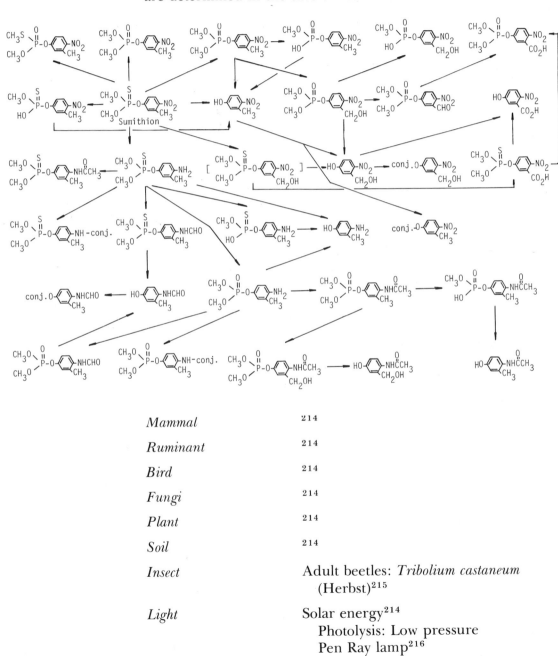

Mammal	214
Ruminant	214
Bird	214
Fungi	214
Plant	214
Soil	214
Insect	Adult beetles: *Tribolium castaneum* (Herbst)[215]
Light	Solar energy[214] Photolysis: Low pressure Pen Ray lamp[216]

CYANOX

The characteristic difference between the metabolisms of cyanox and that of other organophosphorus pesticides is the liberation of hydrogen cyanide as a toxic metabolite in photolysis. Liberation of hydrogen cyanide has also been observed in the case of surecide.

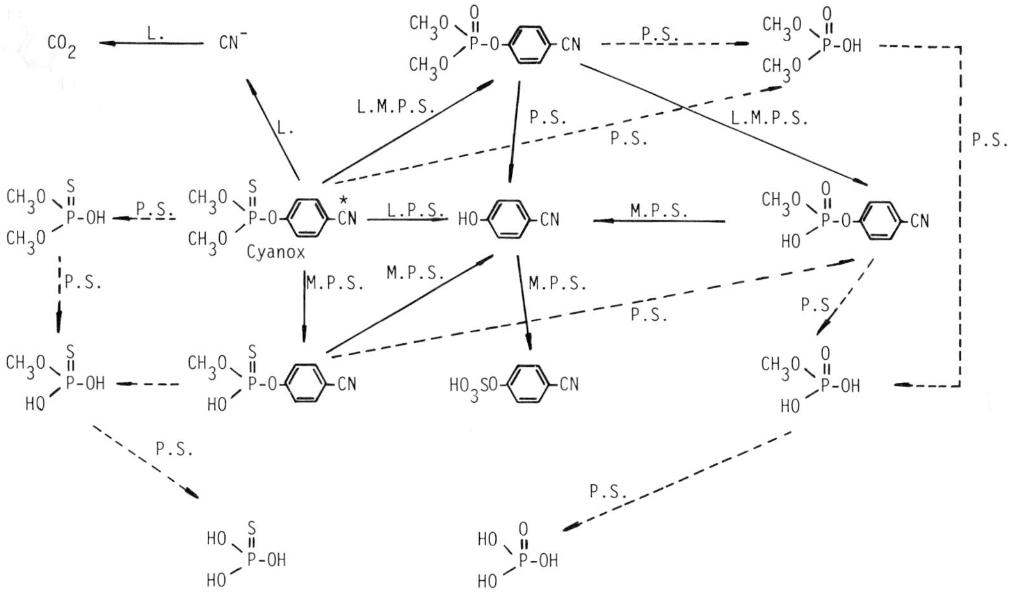

Plant	*Phaseolus vulgaris* L. 'Nagauzura'
Soil	Loam soil (pH 6.5): Setagaya top soil[187]
Light	High pressure mercury vapor lamp[218]
Mammal	Male Wistar rats[217]

BROMOPHOS

Bromophos is oxidized to bromoxon and hydrolyzed to monodemethylbromophos, dimethyl thionophosphate, and 4-bromo-2,5-dichlorophenol in plants.

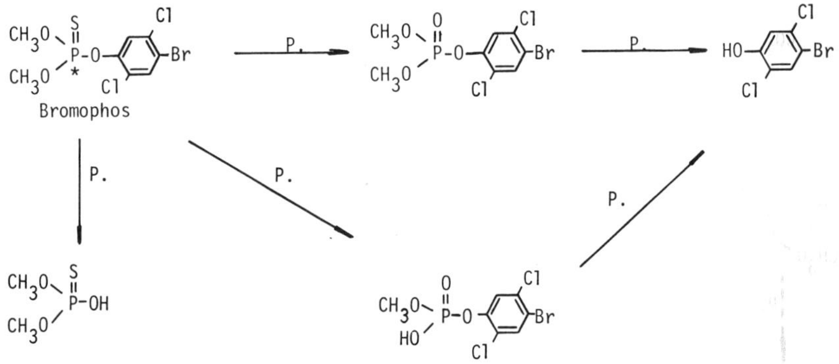

Plant Tomatoes[219]

DASANIT

Dasanit has two easily oxidizable groups, P=S and O=S(CH_3), which are predominantly oxidized to oxon and sulfone, respectively. The structures of these metabolites might be resistant to further degradation in plants and milk.

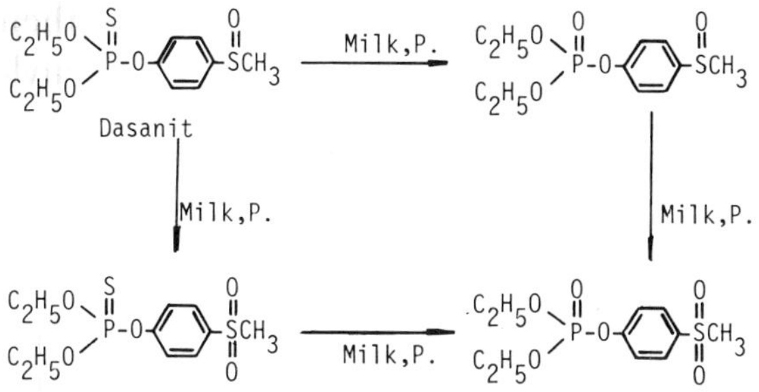

Plant Corn; Grass[220]

Milk 220

ABATE

The P=S group and sulfur atom in the sulfide group of abate undergo stepwise oxidation. The oxidative compounds are then hydrolyzed, and the final metabolite is 4,4′-dihydroxydiphenylsulfone.

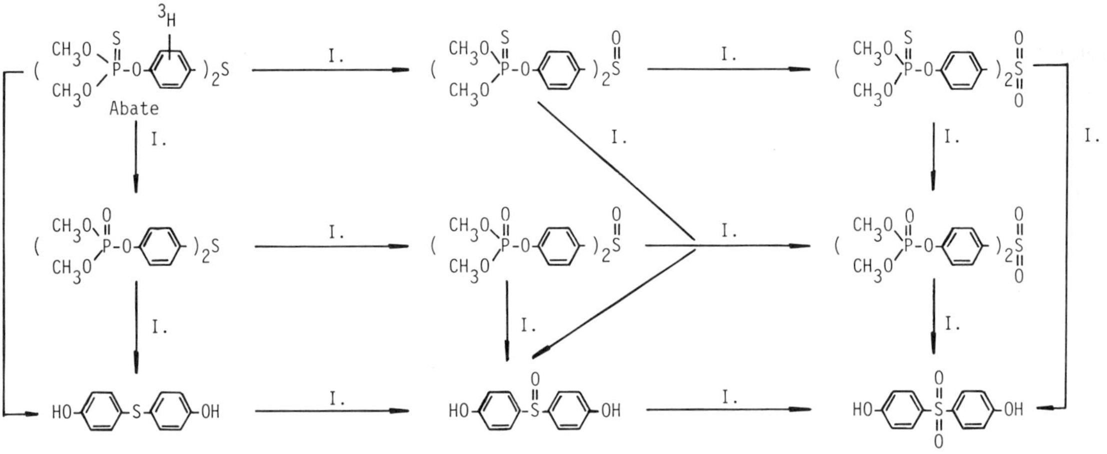

Insect Susceptible 3-day-old female houseflies (*Musca domestica* L. SNAIDM) Fourth-instar mosquito larvae (*Aedes aegypti* L.)[221]

ISOXATHION (KARPHOS)

3-Hydroxy-5-phenylisoxazole, resulting from the rapid cleavage of the phosphonic acid linkage of isoxathion, undergoes glucosidation to form 3-(β-D-glucopyranosyloxy)-5-phenylisoxazole and 2-(β-D-glucopyranosyl)-5-phenylisoxazolin-3-one. The latter is oxidized to the principal metabolite, 2-(β-D-glucopyranosyl)-5-(*p*-hydroxyphenyl)-isoxazolin-3-one. Hydroxylation of the former glucoside might also occur among unidentified minor products. Another metabolic pathway of 3-hydroxy-5-phenylisoxazole via the formation of benzoic acid to its conjugates is negligible, and the oxazole skeleton is quite stable in plants.

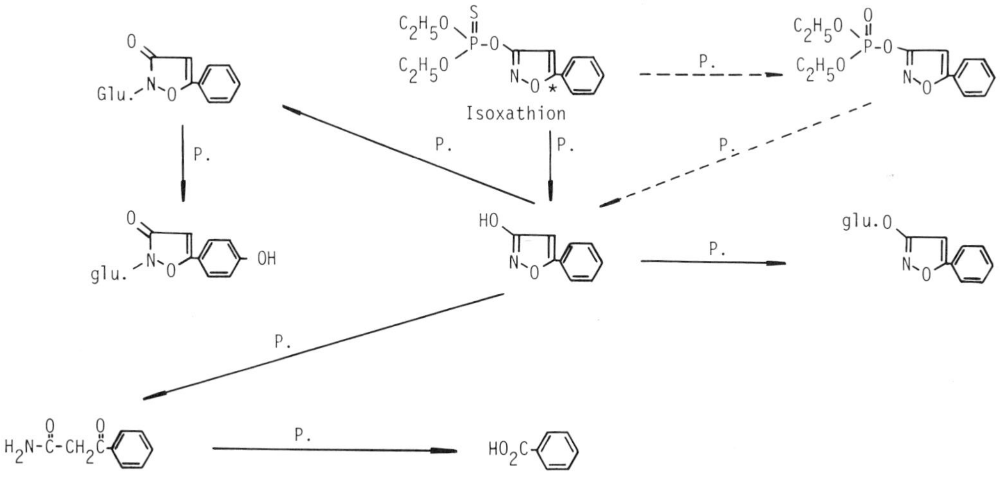

Plant Bean; Cabbage; Chinese cabbage[222]

CHLORPYRIFOS

Oxidative desulfurization of chlorpyrifos gives its oxone and via oxidation and hydrolysis, trichloropyridole is produced. Although dechlorination of chlorpyrifos has not been reported in plants and rat and fish metabolisms, reductive dechlorinated 3-dechlorochlorpyrifos has been identified in insects.

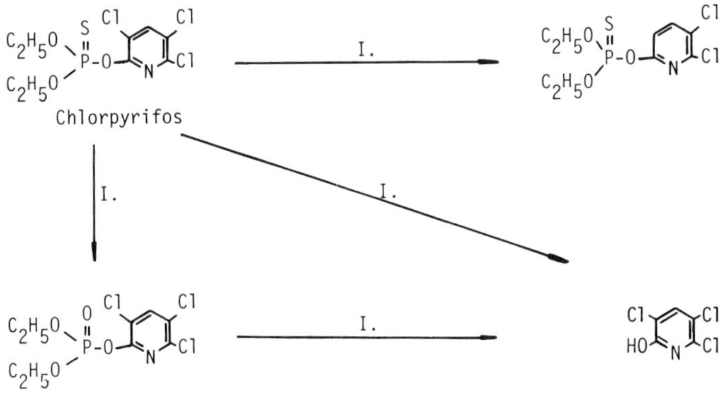

Insect Two hundred termites = Easter subterranean termites: *Reticulitermis flavipes* (Kollar)[223]

DIAZINON

In addition to the oxidative and hydrolyzed products of diazinon, the hydroxylated products of alkyl groups in the pyrimidine ring have been determined to be different types of metabolites in mammals and microsomal systems. Hydroxylation proceeds at both the methyl and isopropyl groups in mammals when phosphate linkages are intact, but in microsomes, only at the isopropyl group after hydrolysis of the ester linkage to form 2-(1-hydroxymethylethyl)- and 2-(1-hydroxy-1-methylethyl)-4-methyl-6-pyrimidones.

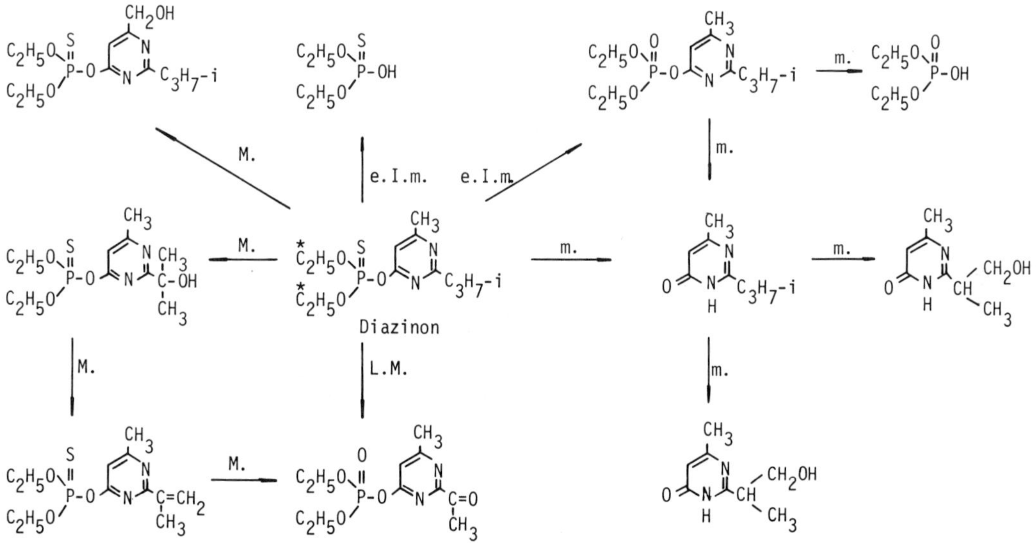

Light	UV irradiation[224]
Microsome	Rat liver enzyme[225]
Insect	Housefly and its enzyme assay[226]
Mammal	Sheep[227,228,229] Rats[229]

PHOXIM

Phoxim is usually degraded to phoxim-oxon by oxidative desulfurization, and to the corresponding oxime and phosphate by hydrolysis. Phoxim is isomerized to phosphorothiolate ester by another metabolic degradation reaction.

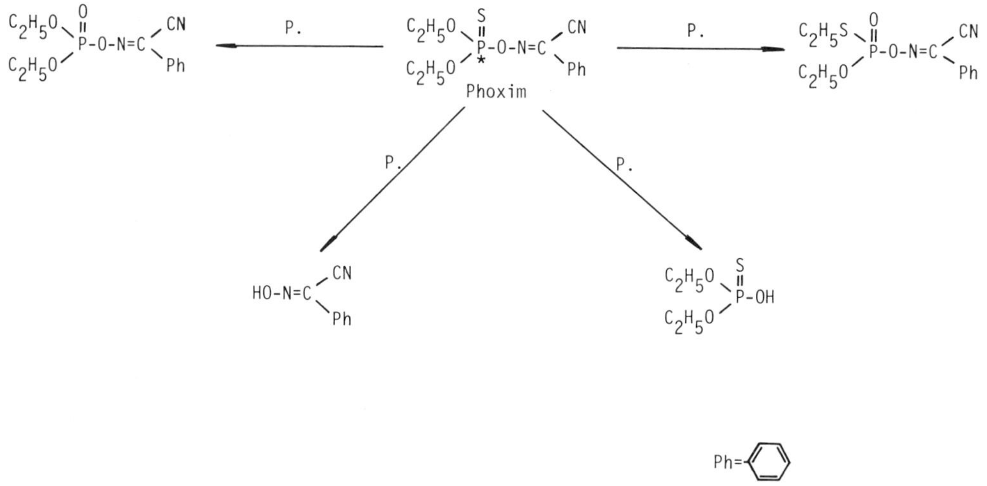

Plant Coastal Bermuda grass; Forage corn: DeKalb-805[230]
Hard red winter wheat[231]

SALITHION

Cyclic linkage of this cyclic phosphate ester, Salithion is easily cleaved by hydrolysis. The resulting product is demethylated from P-*O*-Me. The demethylated products (thion and oxon types) are hydrolyzed to the same product or saligenin, which is then metabolized as conjugates in mammals and plants.

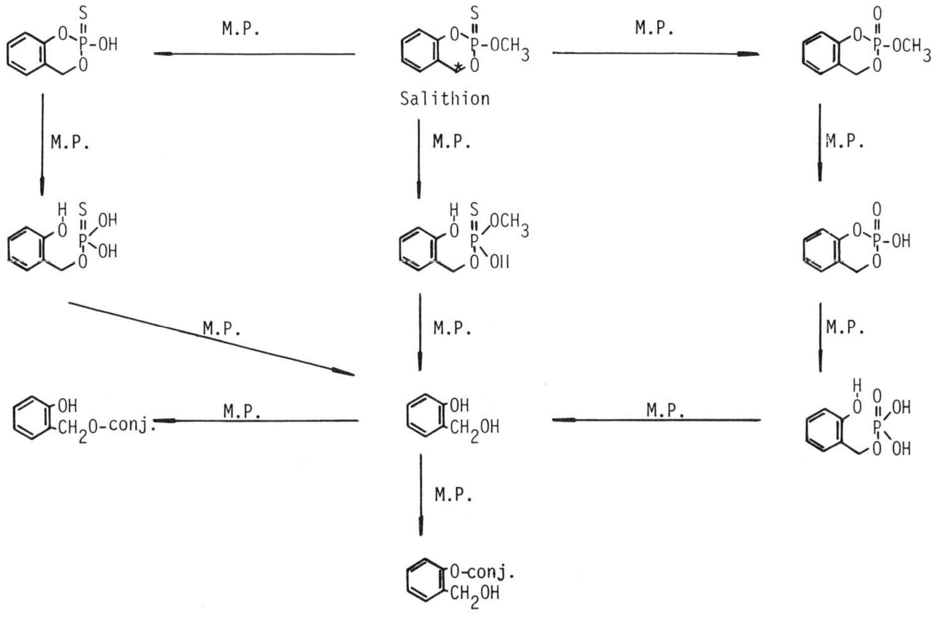

Mammal	Male S-D rats[232]
Plant	Rice: *Oryza sativa;* Bean: *Phaseolus vulgaris* L.[232]

22

Phosphorodithiolates

HINOZAN (EDIFENFOS)

Hydroxylation of hinozan takes place only in rice brast fungus, but trans-esterification or intramolecular exchange of *S*-phenyl and *O*-ethyl radicals is characteristic in rice plants. Further degradation to very polar metabolites of *O*-ethyl *S*-phenylhydrogen phosphorothiolate and phenyldihydrogen phosphorothiolate occur predominantly in fungi and plants, respectively.

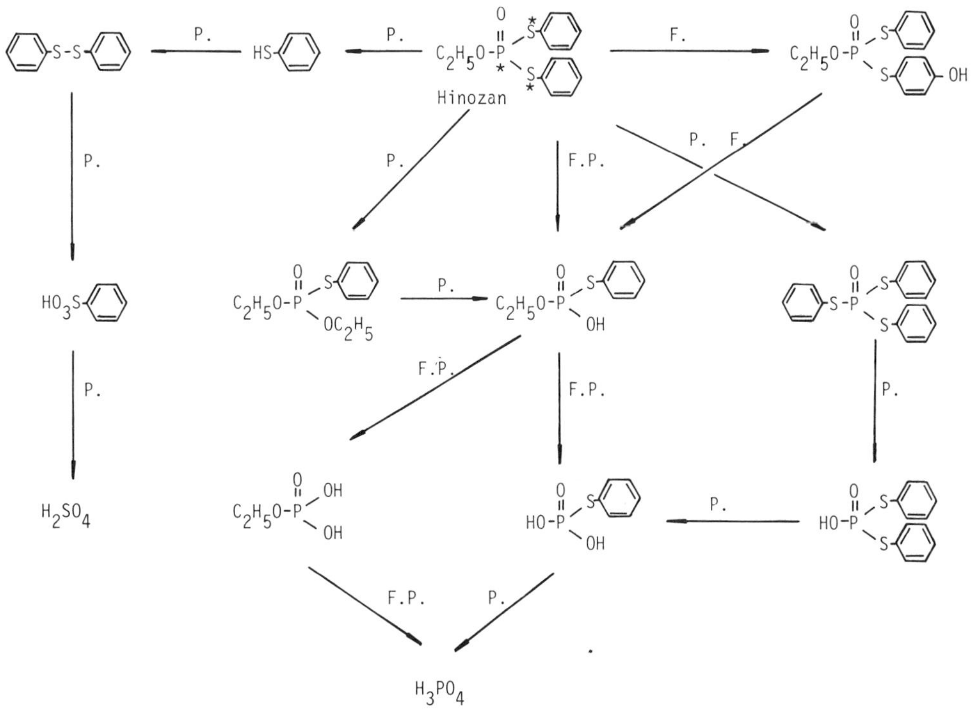

Fungus	Mycerial cells of *Pyricularia oryzae*[233]
Plant	Rice: 'Asahi' variety[234]
	Rice: *Oryza sativa* L. var. Hatsuhinode[235]

MOCAP

Propylthiolate ion liberated from mocap is strongly nucleophilic. Most of the metabolites formed as the result of dealkylation of mocap have an *S*-propyl group in their structures.

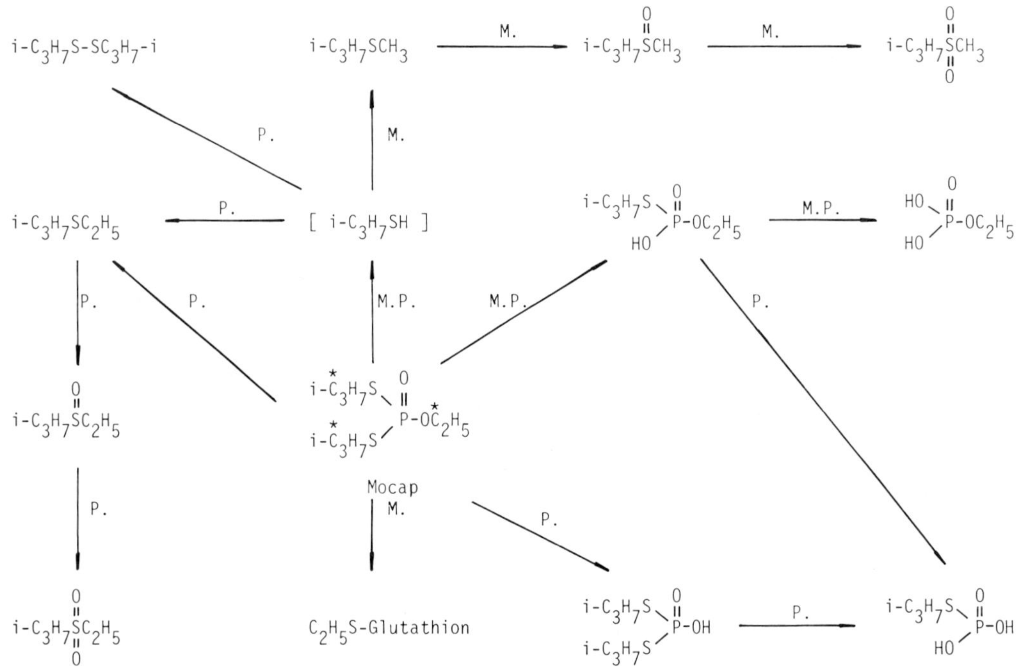

Plant	Snap beans: *Phaseolus vulgaris* L.; Corn: *Zea mays* L.[236]
Mammal	Female and male S-D rats[237]
Microsome	Liver microsomal system of rats and rabbits[237]

23

Phosphorodithioates

MALATHION

Malaoxon has been determined to be the major metabolite at about 25% of the total and is produced mainly by enzymatic oxidation. Predominant degradation is caused mainly by hydrolysis in insects.

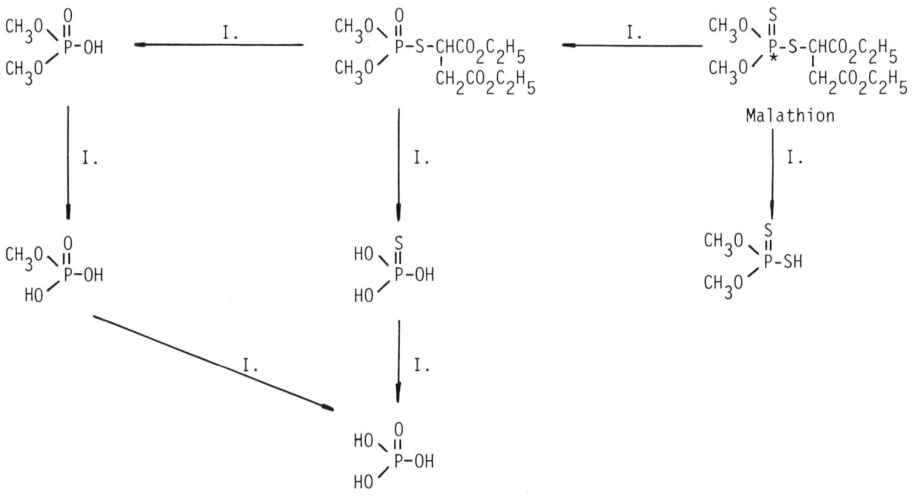

Insect Cotton leaf worm: *Spodoptera littoralis*[238]

FORMOTHION AND DIMETHOATE

Formothion metabolism differs significantly from that of dimethoate in the immediate hydrolytic attack on the N–CHO bond, leading to approximately equal amounts of dimethoate and carboxylic acid. Dimethoate formed from deformylation of formothion is hydroxylated at the *N*-methyl group and is oxidized to thiolate and decomposed to polar metabolites. Biological activity of formothion, dimethoate, and their metabolites in houseflies (LD_{50} and AchE pI_{50}) are given in the references.

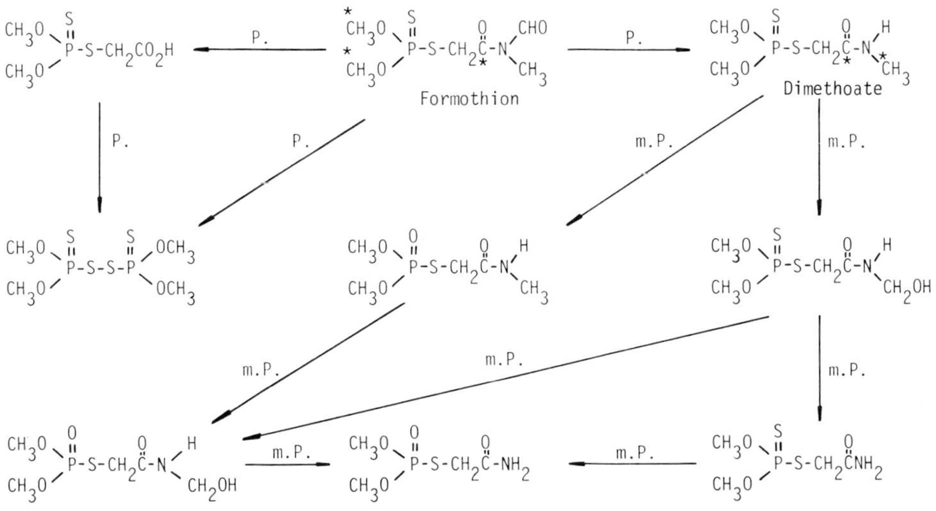

Mammal	Female and male S-D rats[239]
Microsome	Rabbit and rat liver microsomes[237]
Plant	Bean: *Phaseolus vulgaris* L.[239] Bean: *Phaseolus vulgaris* var. Ohnegleichen[240]

BAY NTN 9306

Two oxidation products of BAY NTN 9306 are easily formed. One is the oxon and the other the sulfone, typically observed in these types of esters and substituents.

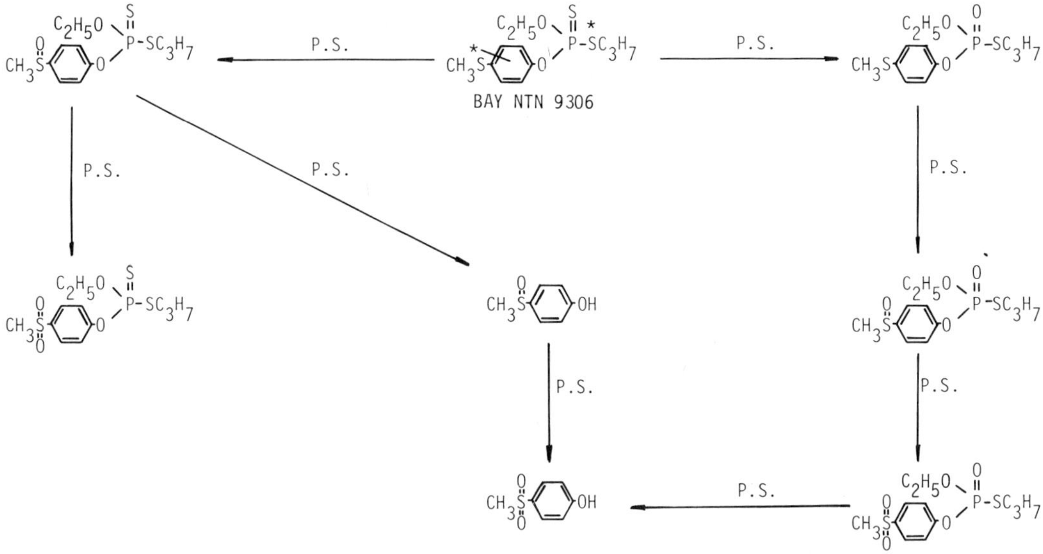

Plant Cotton: Stoneville 213 variety[241]

Soil Lufkin fine sandy loam[241]

SUPRACIDE (METHIDATHION)

5-Methoxy-1,3,4-thiadiazol-2-one is a terminal metabolite of supracide, which is rather stable in plants and soil. The first oxidative degradation reaction takes place with dithioate going to thiolate.

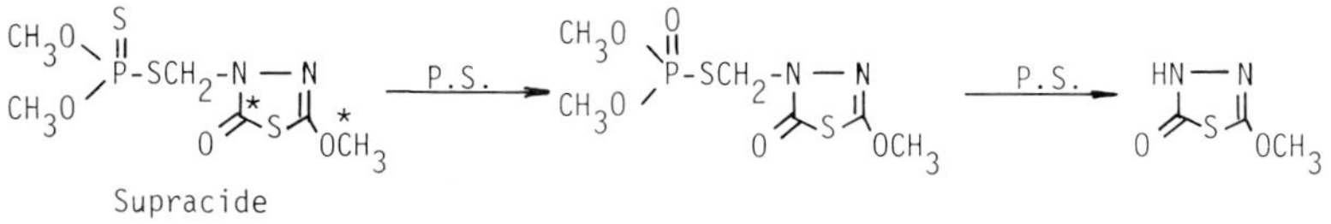

Plant	Alfalfa[242]
	Apples; Cherries; Grapes; Prunes; Potatoes; Hops[243]
Soil	[243]

PHOSALONE

Metabolism of phosalone proceeds by oxidation to oxon, cleavage of the N–CH_2 bond to yield 6-chlorobenzoxazolin-2-one, and then probably by opening of the heterocyclic ring to yield identified 2-amino-7-chloro-3*H*-phenoxazin-3-one.

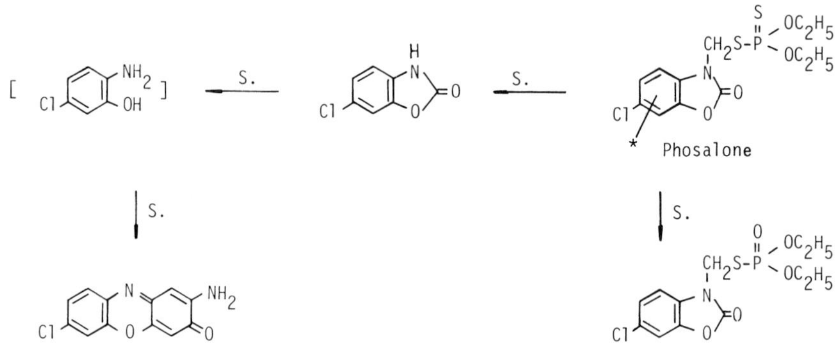

Soil Metapeake loam; Monmouth sandy loam[244]

DIOXATHION

Dioxathion is rapidly metabolized, but ethoxy-containing metabolites other than the oxon, dioxon, and phosphorous acids [$HSP(S)(OC_2H_5)_2$, $HSP(O)(OC_2H_5)_2$, $HOP(O)(OC_2H_5)$] occur in low or trace levels. A small portion of dioxathion might undergo hydroxylation on the alkoxy group and dioxane ring; the latter undergoes hydrolysis to give dialdehyde.

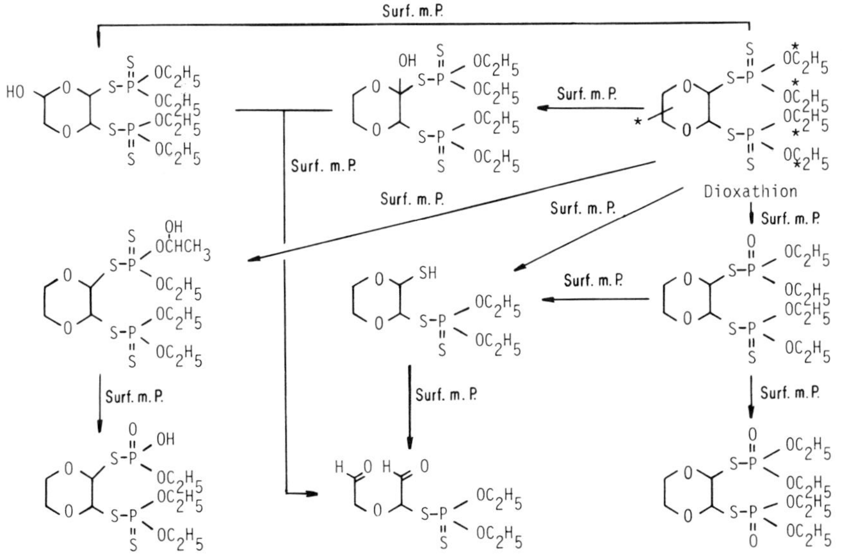

Surface	Glass, silica gel[245]
Microsome	Male albino S-D rat liver microsomes[245]
Plant	Snap bean (Contender var.) leaves[245]

24

Pyrethroids

PYRETHRIN I AND II

Pyrethrin I and II are converted to metabolites which contain a *trans*-2-carboxyprop-1-enyl side chain resulting from oxidation of the crysanthemate isobutenyl group or hydrolysis of the pyrethrate methoxycarbonyl group. The *cis*-2′,4′-pentadienyl side chain of pyrethrin I or II is oxidized at the penta-2,4-dienyl group to give the *cis*-4′,5′-dihydroxypent-2′-enyl derivative, a 4′ conjugate of this diol, or a *trans*-2′,5′-dihydroxypent-3′-enyl derivative.

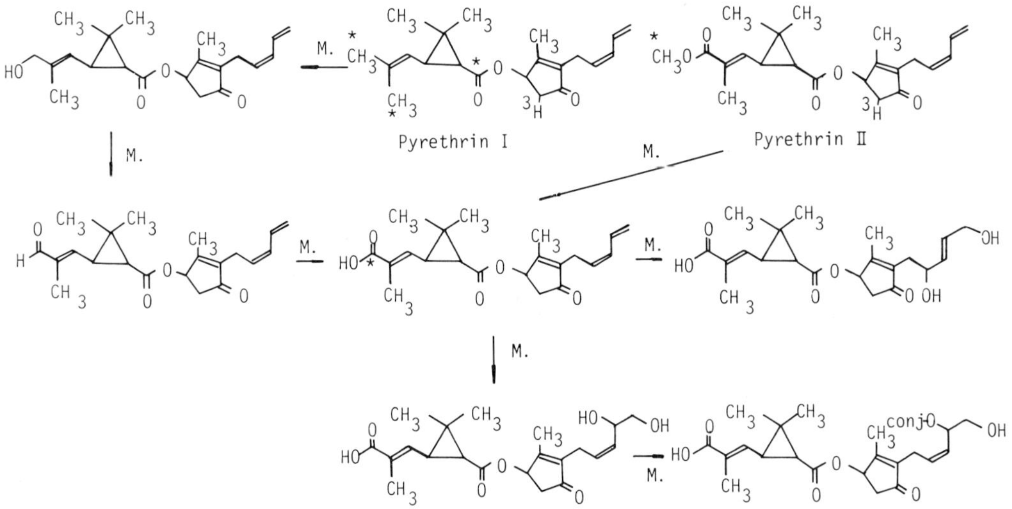

Mammal Male albino S-D rats[246]

ALLETHRIN

Allethrin is oxidized not only at the chrysanthemate isobutenyl moiety to the corresponding primary alcohol, but also at the allyl group to 1′-hydroxyprop-2′-enyl and 2′,3′-dihydroxypropyl derivatives, or at a methyl group on the cyclopropyl moiety to a hydroxy derivative. Allethrin is also converted to chrysanthemum dicarboxylic acid and allethrolone.

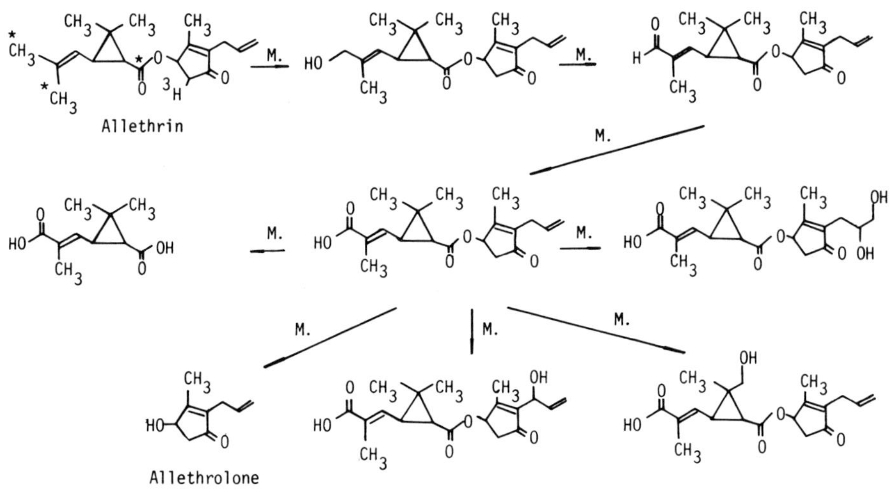

Mammal Male albino S-D rats[246]

(+)-*TRANS*- AND (+)-*CIS*-RESMETHRINS

Some degradative products of the resmethrins are 5-benzyl-3-furoic acid, chrysanthemic acid, the intermediary alcohol and aldehyde oxidation products, and conjugates of each of these acids. The isobutenyl moiety is oxidized at either the *cis*- or *trans*-methyl group in (+)-*cis*-resmethrin, but only at the *trans*-methyl group in (+)*trans*-resmethrin. An unanticipated metabolic pathway involves epimerization at C-3 of the cyclopropane ring. Some metabolites of (+)-*trans*-resmethrin are more toxic than the parent compound.

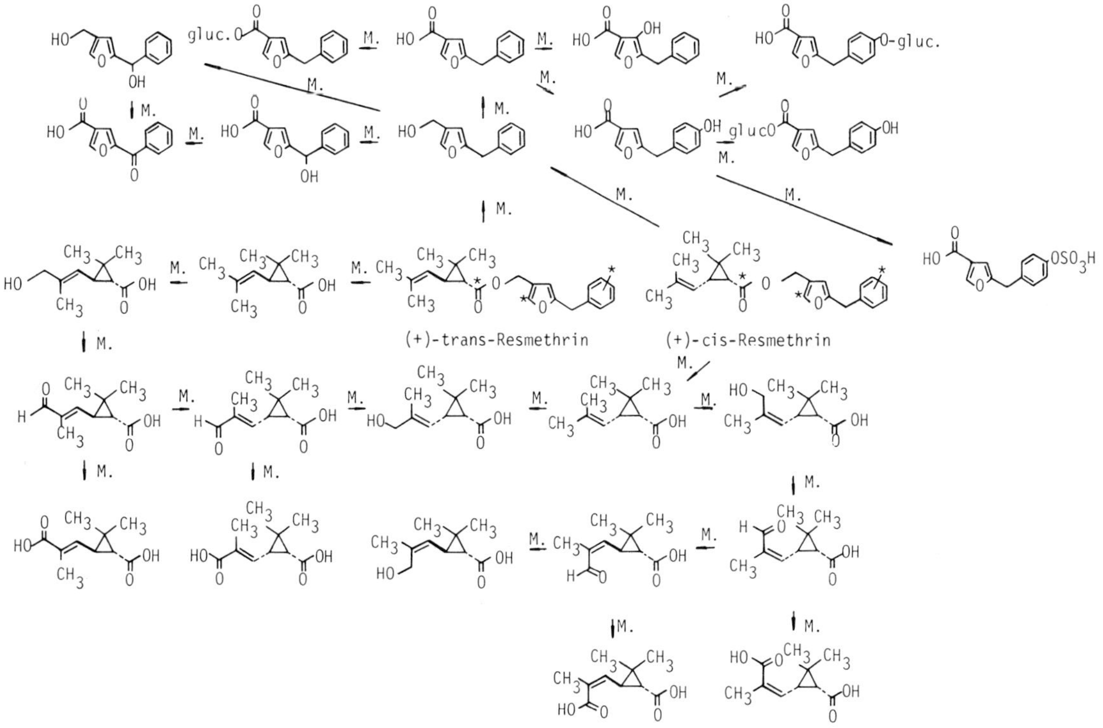

Mammal Male S-D rats[247]
Male albino S-D rats[248]

CIS- AND *TRANS*-CYPERMETHRINS*

The major degradation pathway of cypermethrin is hydrolysis of the ester linkage to give 3-phenoxybenzoic acid and 3-(2,2-dichlorovinyl)-2,2-dimethylcyclopropanecarboxylic acid. (From the cis-isomer both *cis*- and *trans*-cyclopropanecarboxylic acids are found.) A minor degradative route is ring hydroxylation to give an α-cyano-3-(4-hydroxyphenyl)benzyl ester followed by hydrolysis to produce the corresponding hydroxycarboxylic acid.

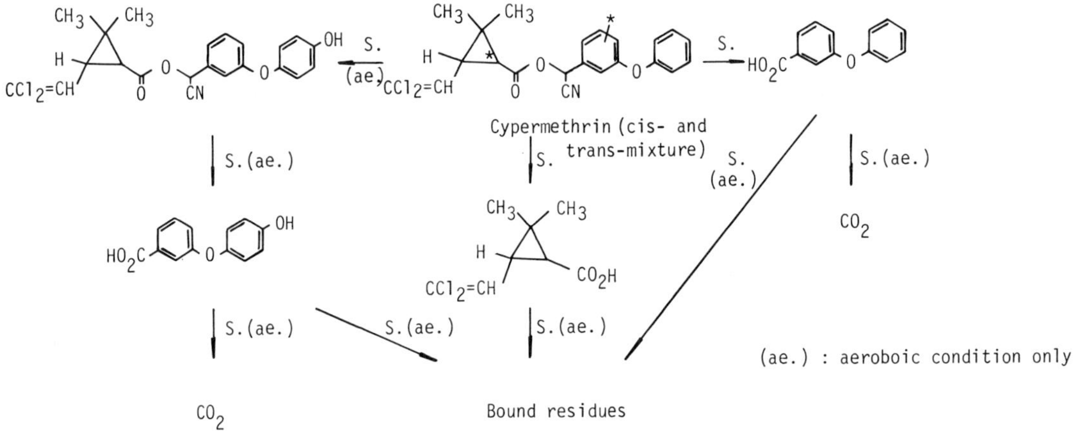

Soil Brenes (sandy clay, pH 8.0) and Los Palacios (clay pH 7.7) soils in Spain; Leiston (sandy loam pH 6.8) in the United Kingdom[249]

DECAMETHRIN

Twenty-five photoproducts have been identified from light irradiation of decamethrin or its identical photolytic derivatives. Photolytic reactions include cis–trans isomerization, ester cleavage reactions, and loss of bromine. Decamethrin undergoes photolysis more rapidly than the two related pyrethroids, NRDC-149 and permethrin. The mixtures of decamethrin photoproducts from solutions and solid-phase reactions have been found less toxic to mice than decamethrin when given intraperitoneally.

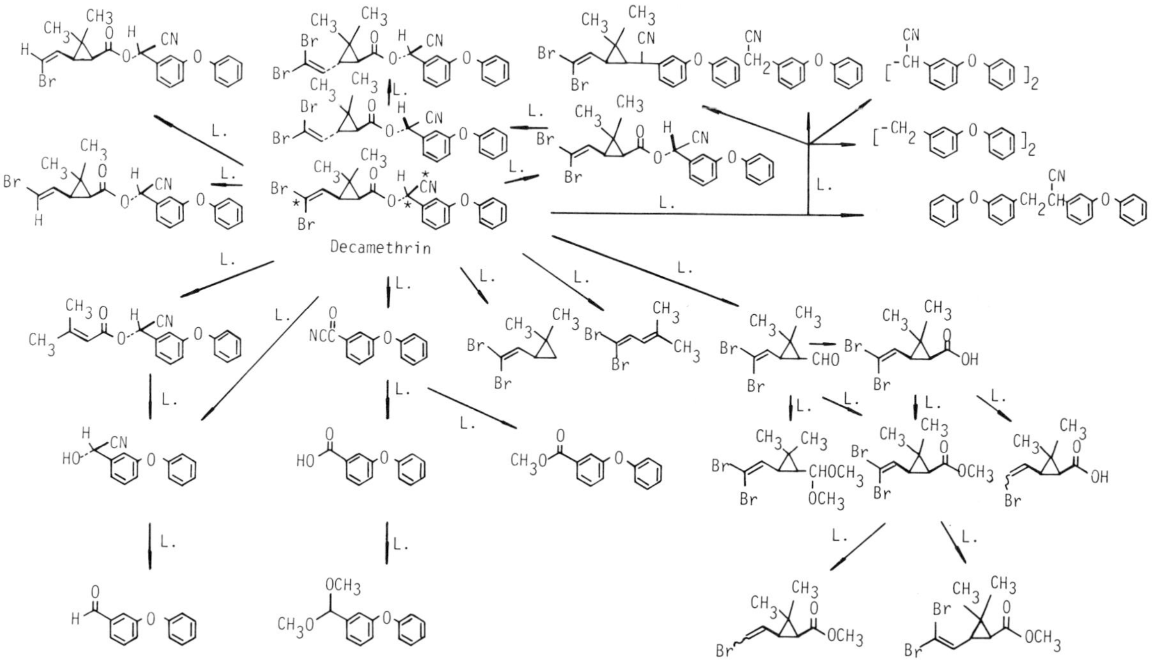

Light Sunlight or UV light irradiation[250]

CIS- AND *TRANS*-PERMETHRINS

cis-Permethrin is more stable than *trans*-permethrin. Both 2′- and 4′-positions of the phenyl rings, and the methyl groups in the cyclopropane rings are metabolized to give some esters hydroxylated at the 2′- and/or 4′- positions of the phenyl rings and methyl groups in the cyclopropane rings. Hydrolysis of permethrin, followed by oxidation, gives the significant metabolites, or phenoxybenzoic acid (free and glucuronide and glycine conjugates), the sulfate conjugate of 4′-hydroxy-3-phenoxybenzoic acids, the sulfate conjugate of 2′-hydroxy-3-phenoxybenzoic acid (from *cis*-permethrin only), the *trans*- and *cis*-dichlorovinyldimethylcyclopropanecarboxylic acid (free and glucuronide conjugates), and the 2-*trans*- and 2-*cis*-hydroxymethyl derivates of each *trans*- and *cis*-acid (free and glucuronide conjugates).

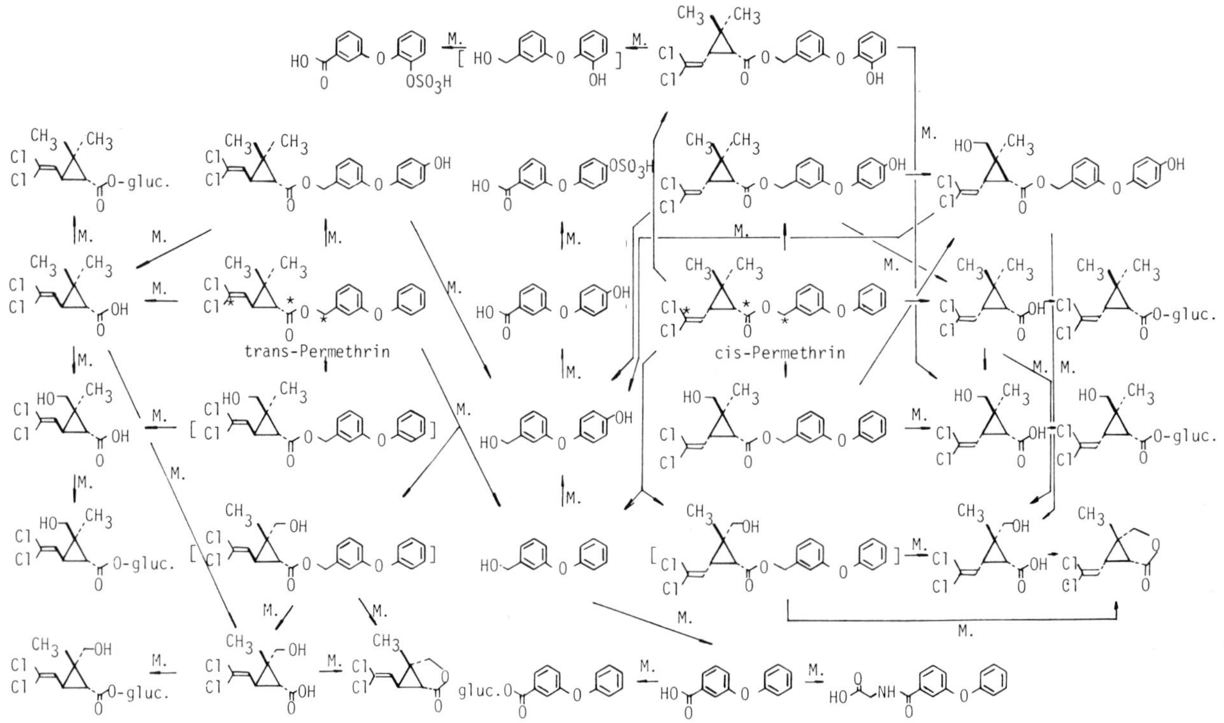

Mammal Male albino S-D rats[251]

FENVALERATE

Fenvalerate undergoes hydroxylation to give 2′- or 4′-hydroxylated phenoxy esters and hydrolysis to give 3-phenoxybenzoic acid and its hydroxy derivatives (free and conjugates), 3-(4-chlorophenyl)-isovaleric acid and its hydroxy derivatives (free, lactones, and conjugates), thiocyanate, and CO_2.

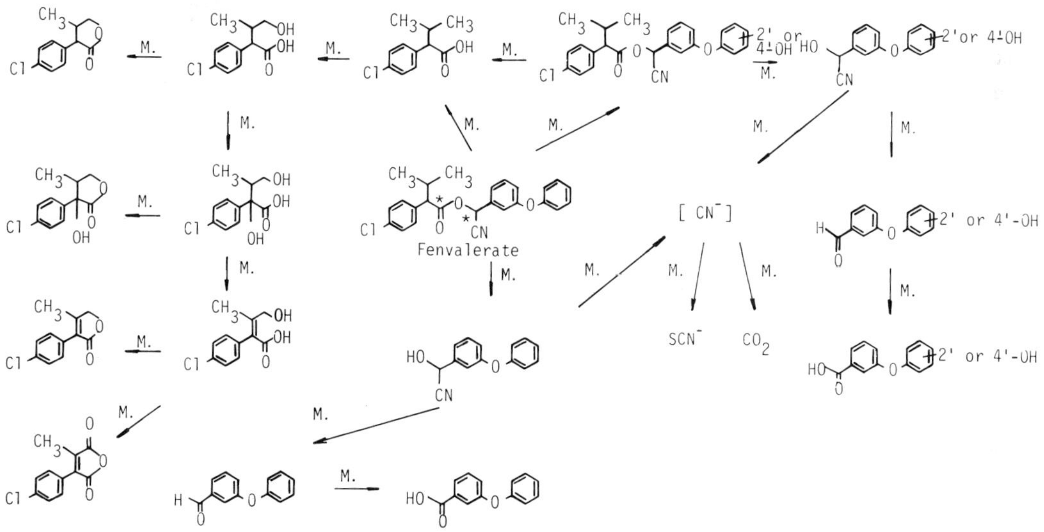

Mammal Male S-D rats[252]

25

Pyridines

CTP (6-CHLORO-α-TRICHLOROPICOLINE)

The α-trichloromethyl group of CTP is oxidized to carboxylic acid and conjugated with glycine in mammals.

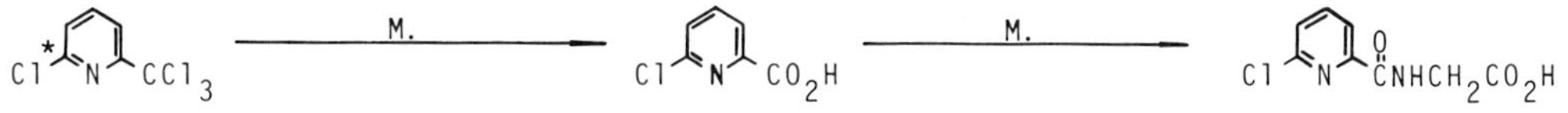

CTP

Mammal Female and male S-D rats[253]

DCP
(3,6-DICHLORO-α-PICOLIC ACID)

One of chlorine atoms at the α-position is replaced by a hydroxyl group and the resulting pyridinol is tautomerized to the α-pyridone analog in soil.

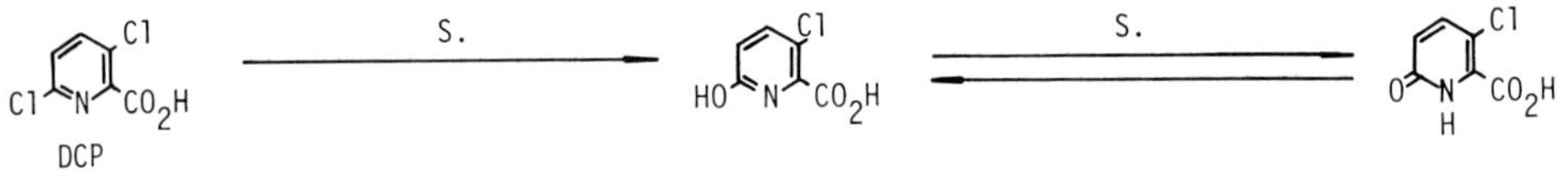

Soil Loam (pH 7.7); Sandy loam (pH 8.1); Luvisol (pH 7.2)[254]

PYRAZON

In light and water, the *N*-phenyl linkage in pyrazon is easily cleaved to the corresponding aminopyridone.

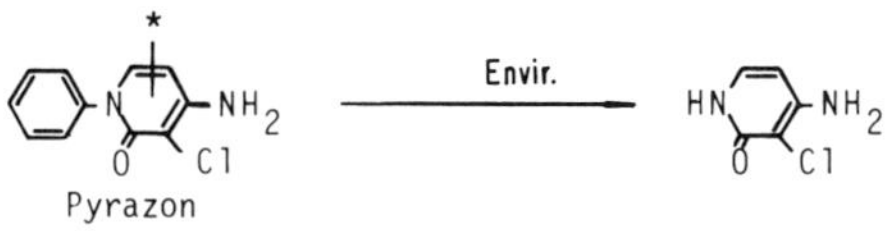

Environment Light and water in model ecosystem[255]

26

Triazines

(*s*-TRIAZINES)

In the metabolism studies of atrazine, bladex, prometone, cyprazine, GS-14254, and other triazines, N-dealkylation of their cysteine and glutathione conjugation is usually observed, including oxidation of *N*-alkyl groups. When an alkoxy group is substituted in the *s*-triazine ring, as in prometone and GS-14254, it is replaced by a hydroxyl group.

Atrazine

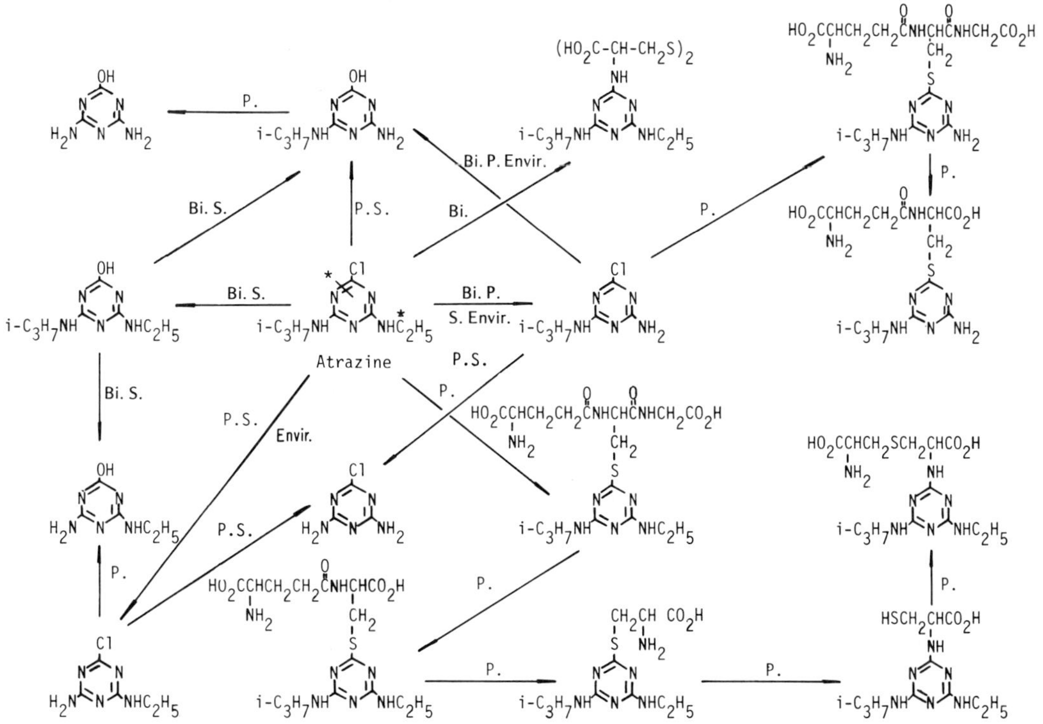

Plant Sorghum: *Sorghum vulgare* Pers., var. North Dakota 104[256,257,258]
Corn: *Zea mays* L., hybrid KE 449; Sugarcane: *Saccharum officinarum* L. hybrid C.P. 61-37 × C.P. 56-59; Barley: *Hordeum vulgare* L., var. Dickson; Oats: *Avena sativa* L., var. Rodney; Pea: *Pisum sativum* L., var. Little Marvel;

	Wheat: *Triticum aestivum* L., var. Chris; Soybean: *Glycine max.* Merr., var. Hawkeye; Carrot: *Daucus carota* L., var. Sativa DC.; Lettuce: *Lactuca sativa* L., var. Great Lakes 659[257] Sorghum seeds: *Sorghum bicolor* L. Moench var. North Dakota 104[259] Oat plant[262]
Bird	Single Comb White Leghorn hens[260]
Environment	Micro ecosystem: *Spartina alterniflora* (a marsh grass commonly known as smooth cordgrass)[261]
Soil	A peach orchard soil[262]

Bladex

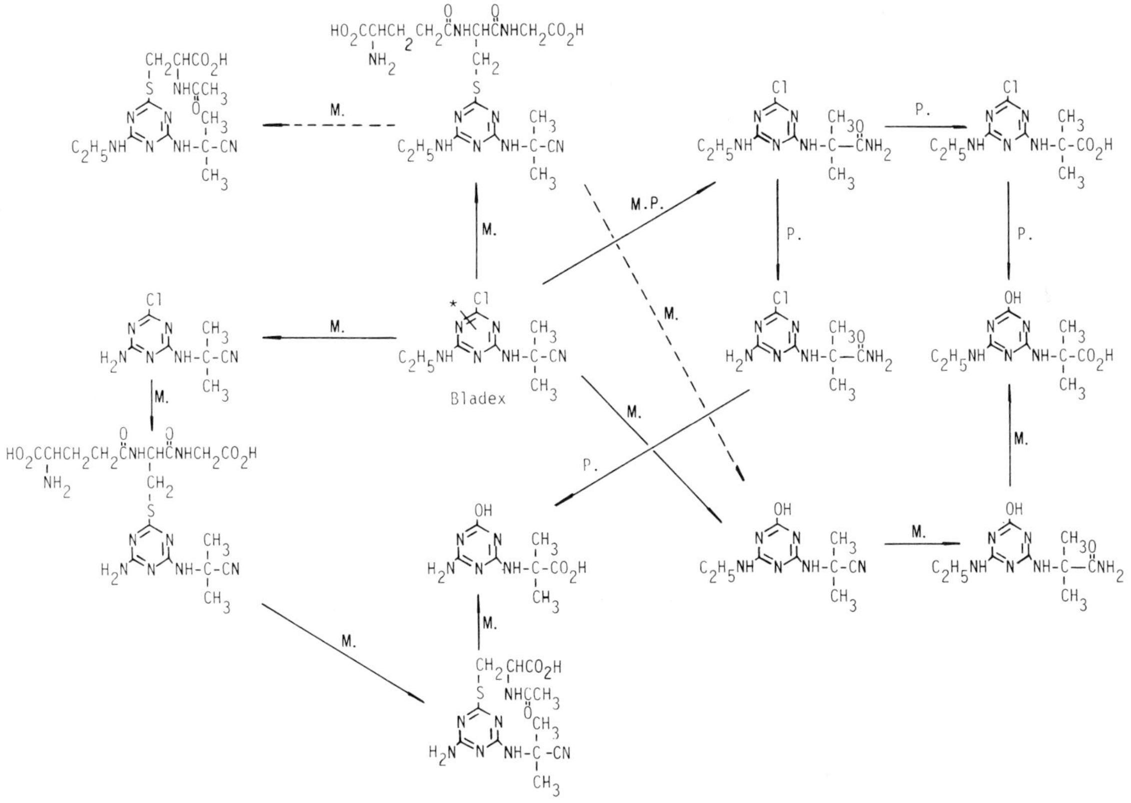

Mammal	Male Carworth Farm E Strain rat[263]
Plant	Crops: Corn silage, stover and ears; Green alfalfa; etc.[264]

Prometone

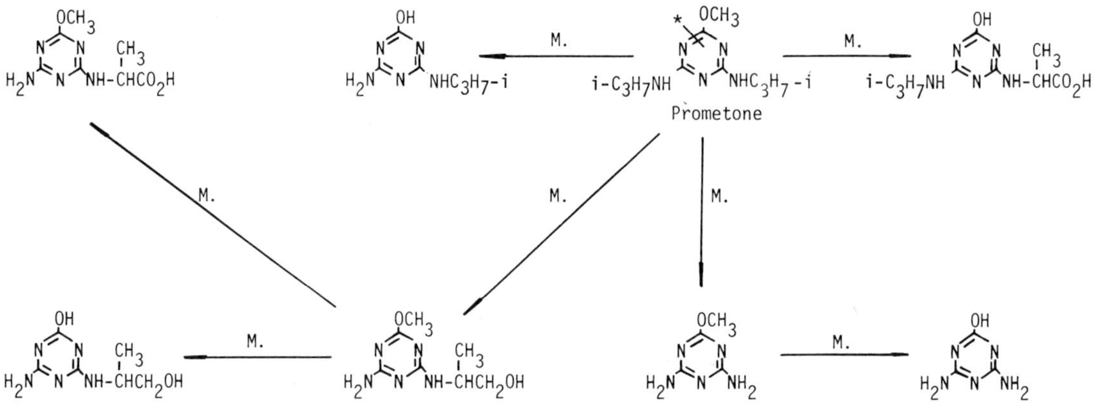

Mammal Rat urine[265]

Cypragine

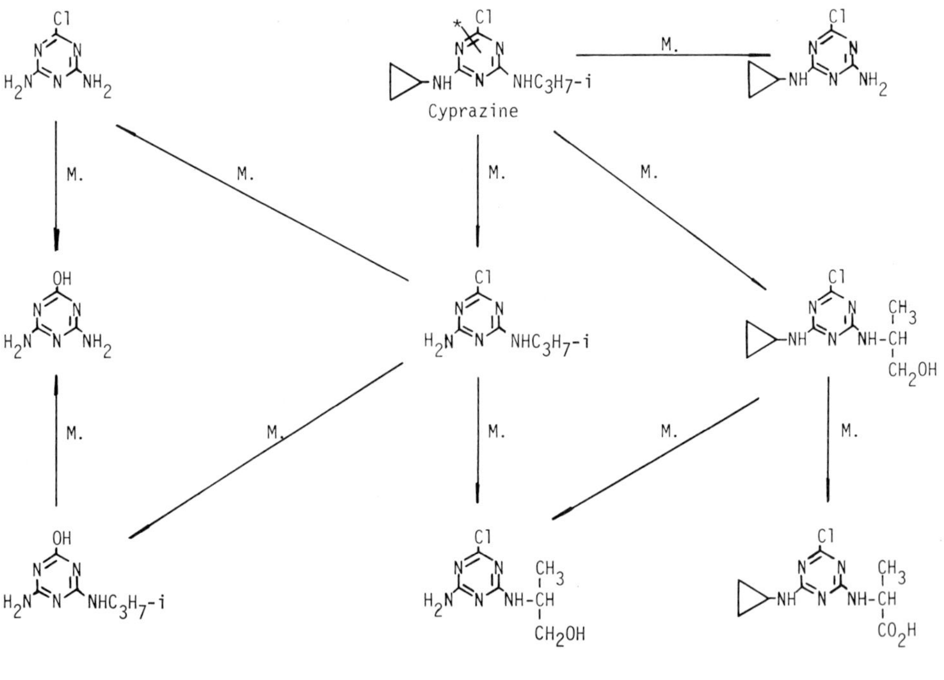

Mammal Rats[266]

GS-14254

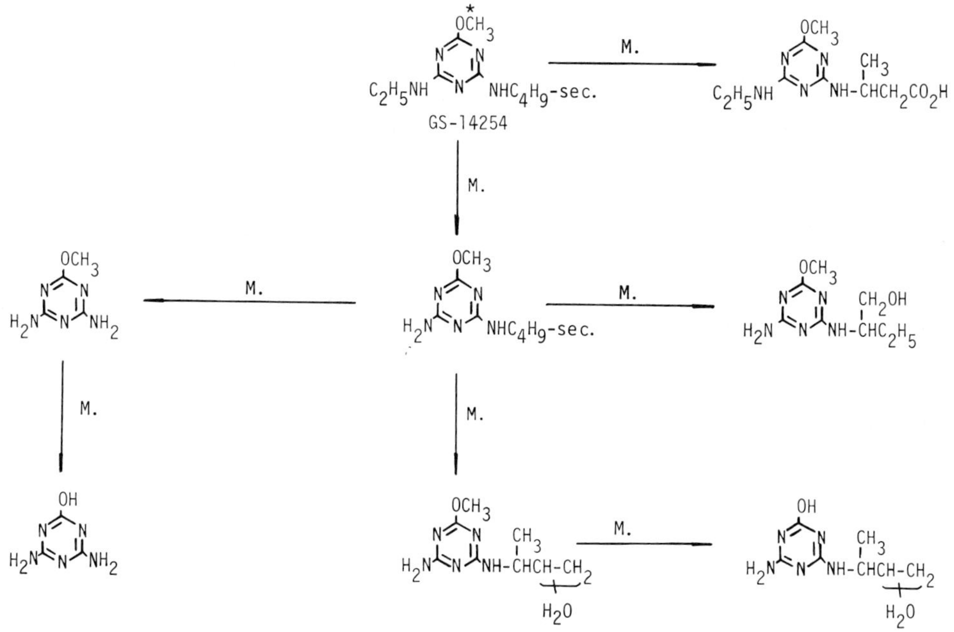

Mammal Lactating dairy cow; Female goat[267]

(*as*-TRIAZINES)

Sencor

The sulfur atom of the S–Me group of sencor is sometimes oxidized to sulfone and is finally eliminated and substituted by a hydroxyl group. This hydroxylated metabolite is equivalent to the keto form, or 4-amino-6-*t*-butyl-1,2,4-triazinyl-3,5-dione. The amino group in the 4-position is further eliminated in plants and soil.

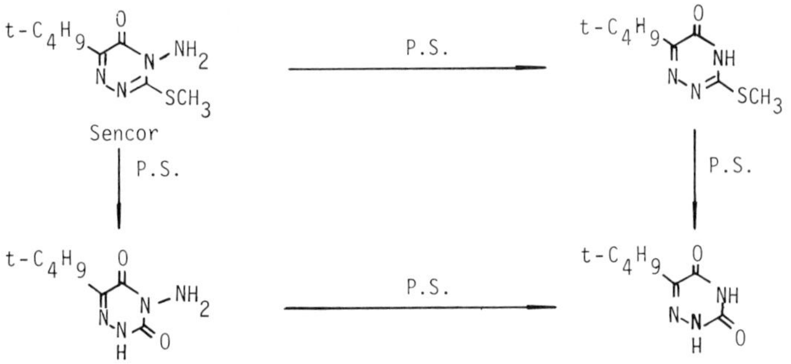

Plant Alfalfa; Beans; Cereal grain; Cereal straw; Sugarcane; Potatoes; Tomatoes[268]

Soil 268

THIAZURIL

The sulfur atom undergoes stepwise oxidation to sulfoxide and sulfone, and the halogen-substituted phenyl ring is hydroxylated, but the other phenyl rings substituted with two methyl groups are rather resistant to oxidation, probably because of the steric hindrance of these two methyl groups and the neighboring *as*-triazine rings.

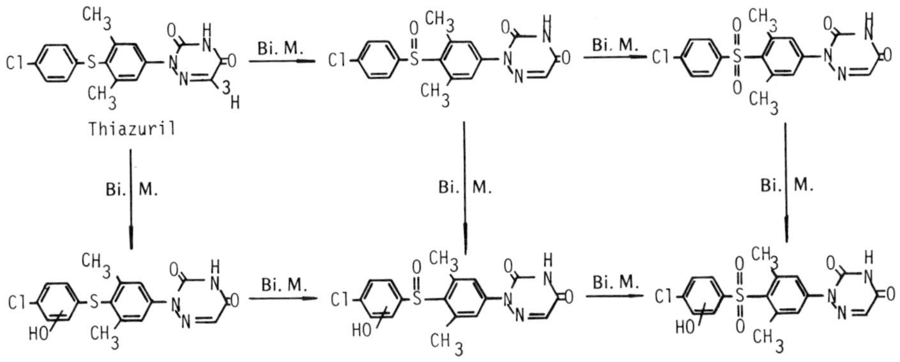

Bird	White Rock broilers[269]
Mammal	Charles River strain rats; Adult beagle dogs (female and male); Male rhesus monkeys[269]

27

Substituted Benzenes and Miscellaneous Compounds

MONOCHLOROACETIC ACID

Monochloroacetic acid is initially metabolized to form *S*-carboxymethylglutathione. This is then converted to *S*-carboxymethylcysteine, part of which is further metabolized to thiodiacetic acid. The other initial metabolism reaction proceeds probably by enzymatic hydrolysis of the carbon–chlorine bond with the formation of glycolic acid in mammals.

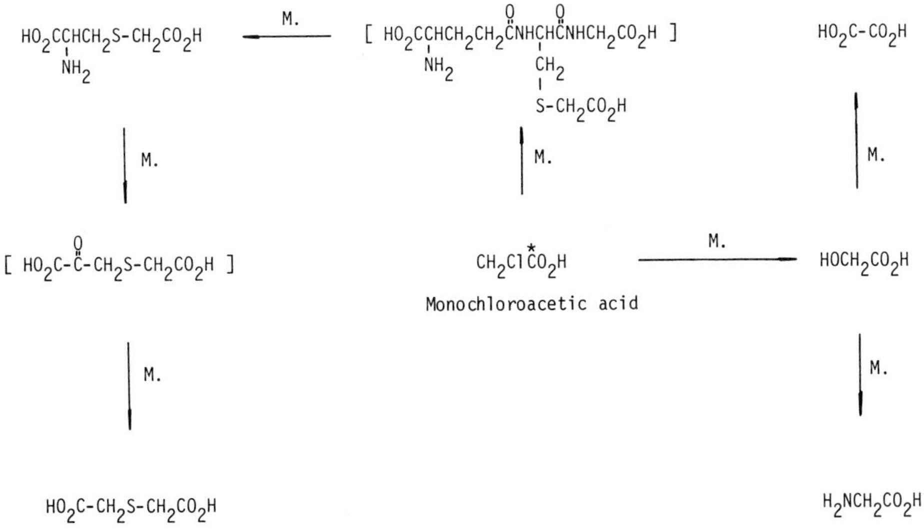

Mammal Female albino mice[270]

2,3,6-TBA (2,3,6-TRICHLOROBENZOIC ACID)

Ring hydroxylation products at the 4- and 5-positions are the initial metabolites of 2,3,6-TBA. These hydroxylated metabolites are then decomposed by decarboxylation and their corresponding phenols are oxidized to catechols in bacteria.

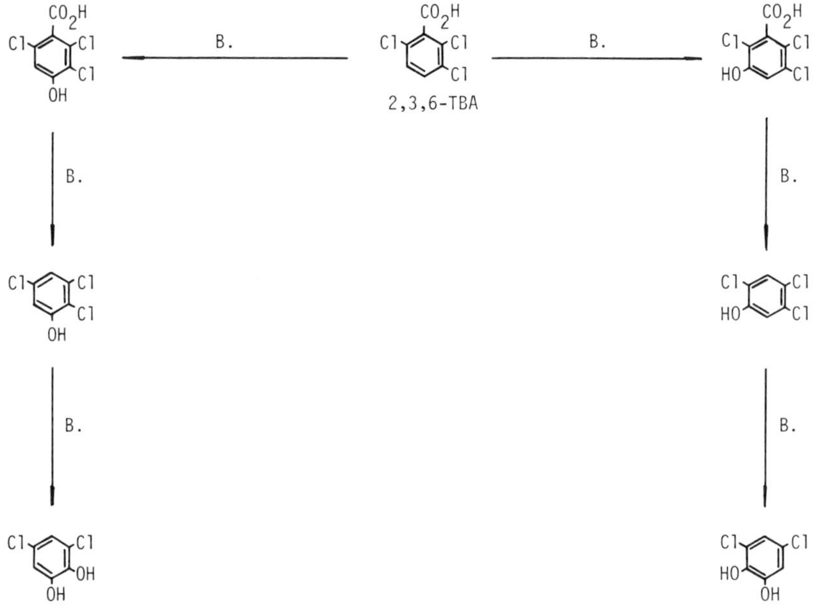

Bacteria *Brevibacterium* sp.[271]

DICHLOROBENIL

Hydrolysis of the nitrile group and hydroxylation at the 3-position of the phenyl ring of dichlorobenil are the main metabolic routes in the initial degradation step, but all routes eventually yield the polar metabolite, 3-hydroxy- 2,6-dichlorobenzoic acid.

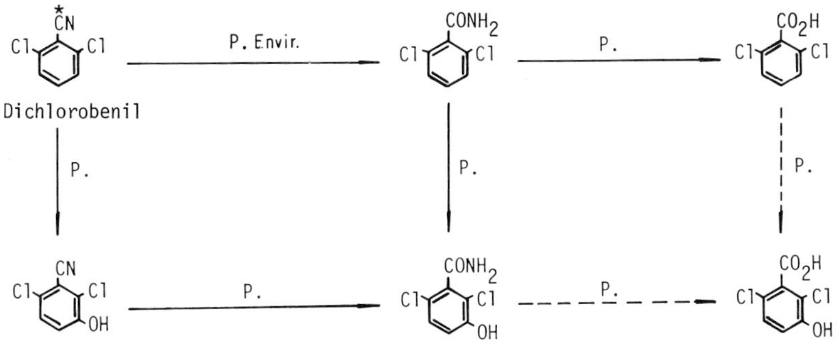

Plant Shoot cutting of alligator weed and parrot feather[272]

Environment Pond water and sediment[273]

DISUGRAN

A major metabolite of disugran is 2-hydroxydisugran; this is further degraded to 2-hydroxydicamba. Herbicidal dicamba is formed by hydrolysis of the methyl ester group of disugran in rumen.

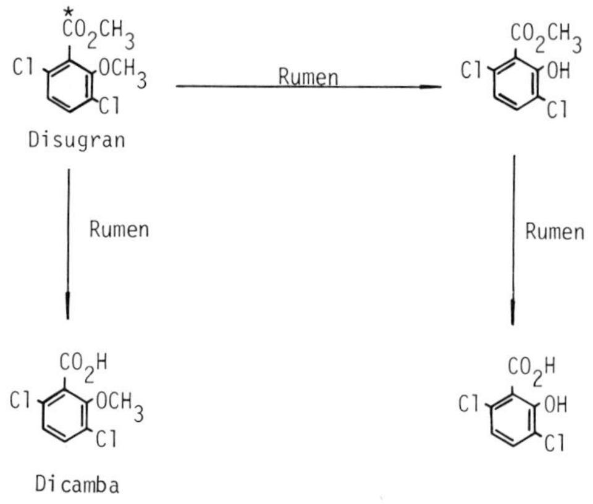

Rumen Rumen fluid of sheep (ewes) mixed breed[274]

CHLORONEB

As the metabolic pathway shows, chloroneb (*p*-hydroquinone dimethyl ether) is metabolized to its corresponding *p*-quinone in plants and soil.

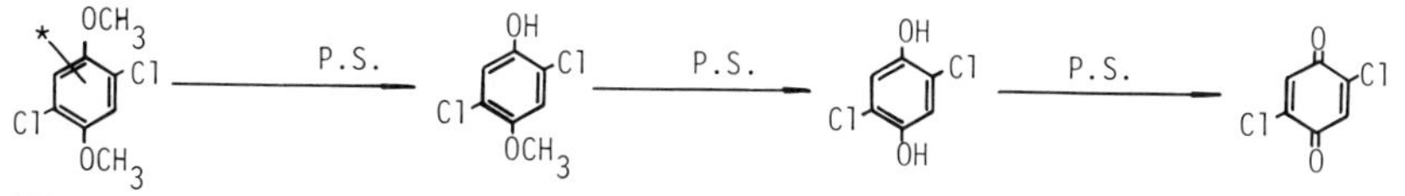

Chloroneb

Plant	Cotton: Stoneville 213 reginned cotton seed; Bean: Wade snap beans[275]
Soil	Keyport silt loam[275]

BROMOXYNIL

Free bromoxynil (2,6-dibromo-4-cyanophenol) results from initial hydrolysis and is followed by three consecutive or concurrent metabolic steps: (a) hydrolysis of the cyano group to amide and to carboxylic acid, followed by decarboxylation to 2,6-dibromophenol, (b) replacement of one or two bromine atoms by a hydroxyl group to 3-bromo-4,5-dihydroxybenzonitrile and 3,4,5-trihydroxybenzonitrile or their hydrolyzed products, and (c) replacement of one or two bromine atoms by hydrogen, giving 3-bromo-4-hydroxybenzonitrile and 4-hydroxybenzonitrile or their hydrolyzed products in plants.

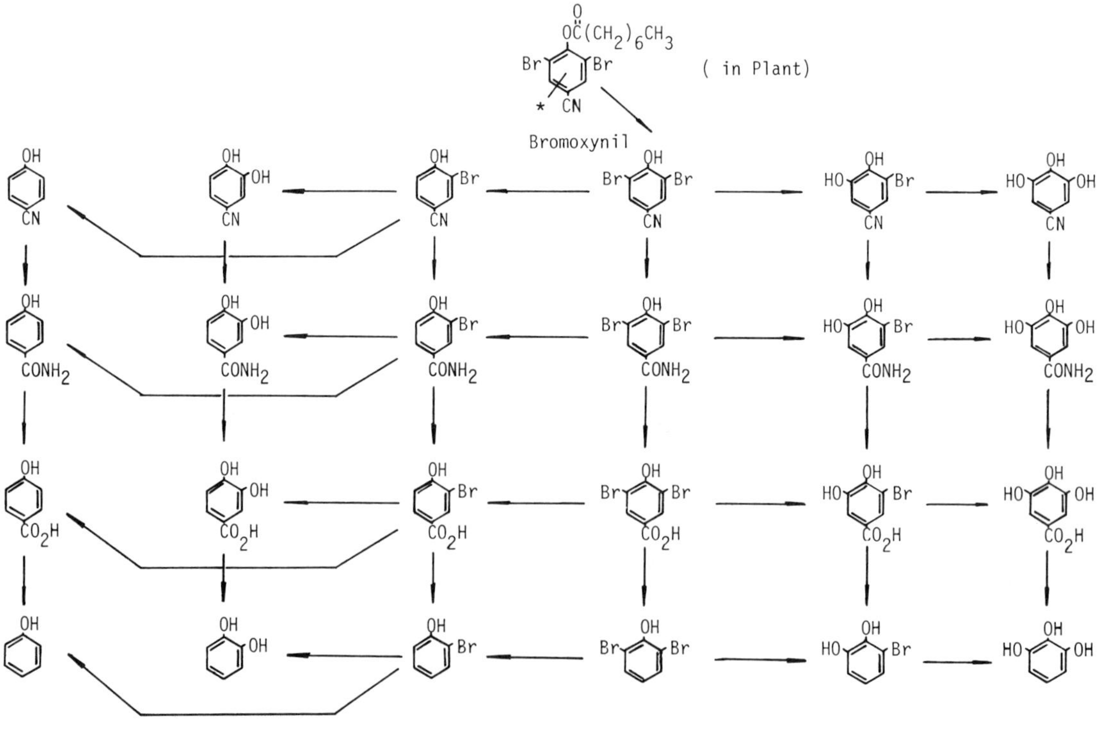

Plant Wheat: *Triticum valgare* var. Kloka[276]

PCP (PENTACHLOROPHENOL)

Various reductive dechlorinations to form trichlorophenols have been observed in the metabolism of PCP. A methyl ether of PCP has been identified and formation of *p*-hydroquinone and its dimethyl ether or cleavage of the phenyl ring to form dicarboxylic acid has been suggested.

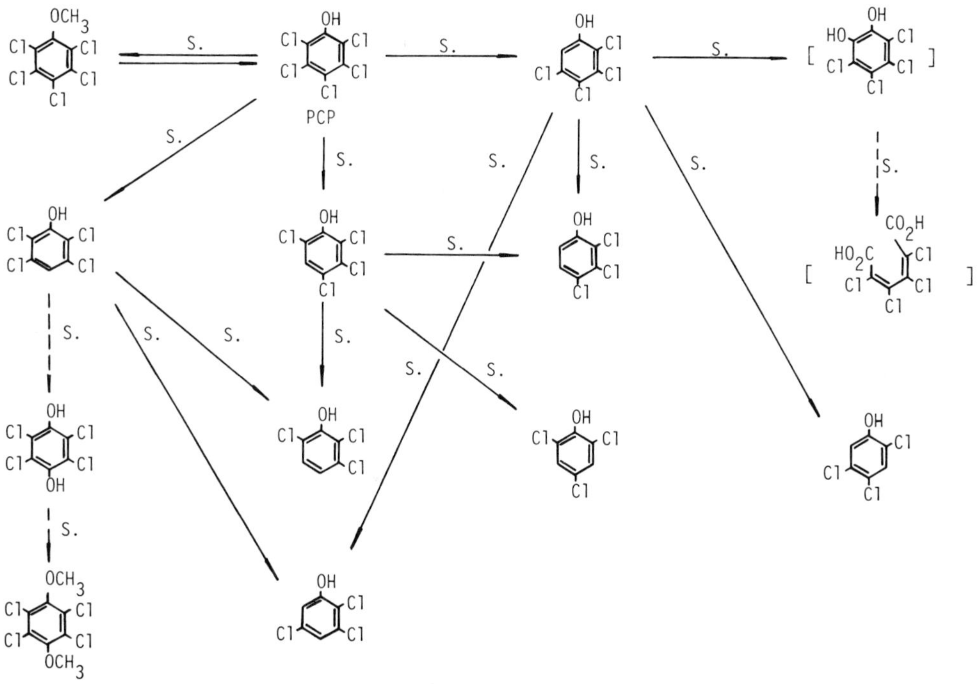

Soil 5

NK-049 (METHOXYPHENONE)

Two methyl and one methoxy groups are substituted separately in the two phenyl rings of the benzophenone of NK-049. Methyl groups are stepwise oxidized as usual to form a hydroxylmethyl and a carboxyl group. The methoxy group is degraded to a hydroxyl group and the final metabolite of NK-049 is 3,3′-dicarboxy-4-hydroxybenzophenone in mammals.

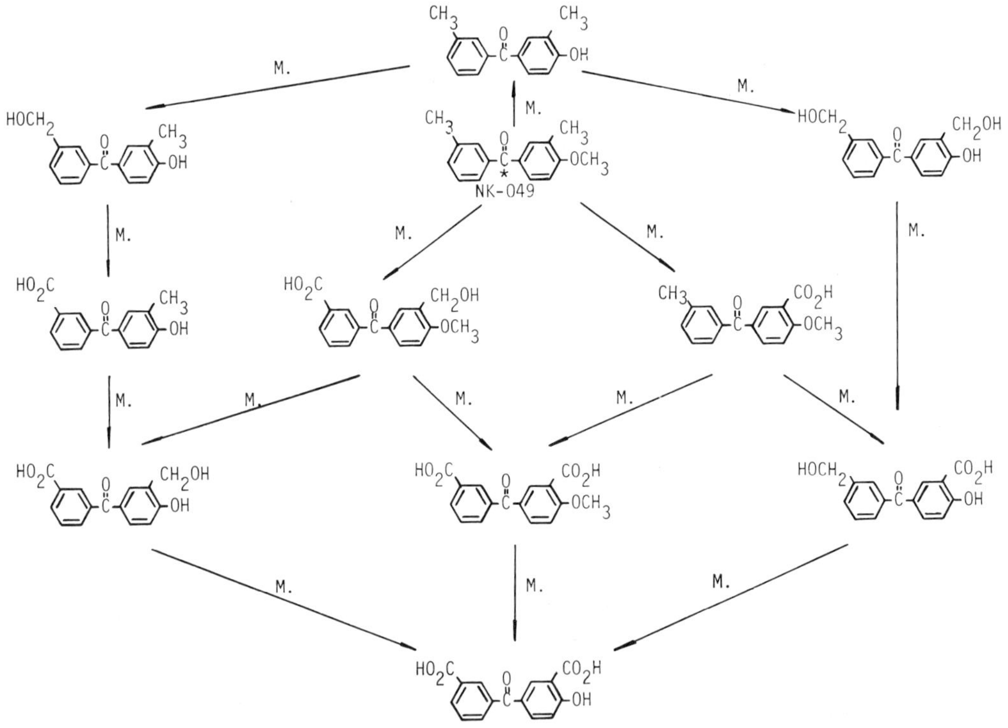

Mammal Female Wistar rats[277]

1-NAPHTHYLACETIC ACID

The 5-position of 1-naphthylacetic acid is enzymatically hydroxylated and the parent compound itself forms its conjugate as an aspartate in plants.

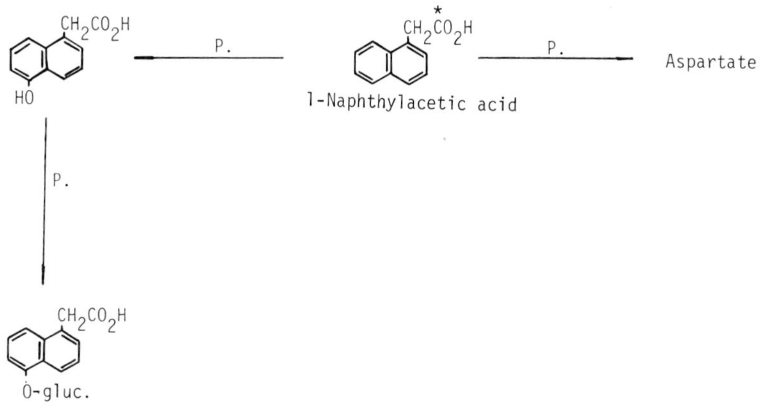

Plant Kinnow mandarin fruiting branch[278]

REFERENCES

1. P. R. Wallnöfer, S. Safe, and O. Hutzinger, *J. Agric. Food Chem.* **21,** 502 (1973).
2. R. Y. Yih and C. Swithenbank, *J. Agric. Food Chem.* **19,** 314 (1971).
3. J. D. Fisher, *J. Agric. Food Chem.* **22,** 606 (1974).
4. R. Y. Yih and C. Swithenbank, *J. Agric. Food Chem.* **19,** 320 (1971).
5. S. Kuwatsuka, *J. Pestic. Sci.* **2,** 201 (1977).
6. A. Warrander and R. Waring, *Pestic. Sci.* **8,** 54 (1977).
7. T. R. Roberts, *Pestic. Biochem. Physiol.* **7,** 378 (1977).
8. G. L. Lamoureux, L. E. Stafford, and F. S. Tanaka, *J. Agric. Food Chem.* **19,** 346 (1971).
9. S. W. Bingham and R. Shaver, *Weed Sci.* **19,** 639 (1971).
10. D. P. Schultz and B. G. Tweedy, *J. Agric. Food Chem.* **19,** 36 (1971).
11. L. F. Krzeminski, B. L. Cox, and A. W. Neff, *Anal. Chem.* **44,** 126 (1972).
12. S. W. Bingham and R. L. Shaver, *Pestic. Biochem. Physiol.* **7,** 8 (1977).
13. L. L. McGahen and J. M. Tiedje, *J. Agric. Food Chem,* **26** 414 (1978).
14. J. M. Tiedje and M. L. Hagedorn, *J. Agric. Food Chem.* **23,** 77 (1975).
15. Y.-L. Chen and C.-C. Chen, *J. Pestic. Sci.* **3,** 143 (1978).
16. P. R. Wallnöfer, M. Königer, S. Safe, and O. Hutzinger, *J. Agric. Food Chem.* **20,** 20 (1972).
17. R. E. Hornish and J. L. Nappier, *J. Agric. Food Chem.* **26,** 1083 (1978).
18. W.-T. Chin, G. M. Stone, and A. E. Smith, *J. Agric. Food Chem.* **18,** 709 (1970).
19. J. P. Rouchaud, J. R. Decallonne, and J. A. Meyer, *Pestic. Sci.* **8,** 65 (1977).
20. S. Darda, R. L. Darskus, D. Eichler, W. Ost, and M. Wotschokowsky, *Pestic. Sci.* **8,** 183 (1977).
21. S. Darda, *Pestic. Sci.* **8,** 193 (1977).
22. G. W. Ivie, *J. Agric. Food Chem.* **23,** 869 (1975).
23. G. D. Paulson, A. M. Jacobsen, and R. G. Zaylskie, *Pestic. Biochem. Physiol.* **7,** 62 (1977).
24. G. L. Lamoureux and L. E. Stafford, *J. Agric. Food Chem.* **25,** 512 (1977).
25. A. K. Sen Gupta and C. O. Knowles, *J. Agric. Food Chem.* **17,** 595 (1969).
26. K. M. Chang and C. O. Knowles, *J. Agric. Food Chem.* **25,** 493 (1977).
27. C. O. Knowles and H. J. Benezet, *J. Agric. Food Chem.* **25,** 1022 (1977).
28. J. Zulalian and P. E. Gatterdam, *J. Agric. Food Chem.* **21,** 794 (1973).
29. J. Zulalian, D. A. Champagne, R. S. Wayne, and R. C. Blinn, *J. Agric. Food Chem.* **23,** 724 (1975).
30. D. D. Kaufman, J. R. Plimmer, and U. I. Klingebiel, *J. Agric. Food Chem.* **21,** 127 (1973).
31. R. D. Minard, S. Russel, and J.-M. Bollag, *J. Agric. Food Chem.* **25,** 841 (1977).
32. E. Anagstopoulos, I. Scheunert, W. Klein, and F. Korte, *Chemosphere* **1978,** 351 (1978).
33. G. D. Paulson, A. M. Jacobsen, R. G. Zaylskie, and V. J. Feil, *J. Agric. Food Chem.* **21,** 804 (1973).
34. P. Moldéus, H. Vadi, and M. Berggren, *Acta Pharmacol. Toxicol.* **39,** 17 (1976).
35. G. G. Briggs and S. Y. Ogilvie, *Pestic. Sci.* **2,** 165 (1971).
36. N. K. Van Alfen and T. Kosuge, *J. Agric. Food Chem.* **22,** 221 (1974).

37. H. C. Newsom and W. G. Woods, *J. Agric. Food Chem.* **21,** 598 (1973).
38. L. E. Olson, J. L. Allen, and J. W. Hogan, *J. Agric. Food Chem.* **25,** 554 (1977).
39. S. K. Bandal and J. E. Casida, *J. Agric. Food Chem.* **20,** 1235 (1972).
40. J. R. Plimmer and U. I. Klingebiel, *J. Agric. Food Chem.* **22,** 689 (1974).
41. P. C. Kearney, J. R. Plimmer, V. P. Williams, U. I. Klingebiel, A. R. Isenssee, T. L. Laanio, G. E. Stolzenberg, and R. G. Zaylskie, *J. Agric. Food Chem.* **22,** 856 (1974).
42. P. P. Williams and V. J. Feil, *J. Agric. Food Chem.* **19,** 1198 (1971).
43. C. J. Soderquist, D. G. Crosby, K. W. Moilanen, J. N. Seiber, and J. E. Woodrow, *J. Agric. Food Chem.* **23** 304 (1975).
44. E. Leitis and D. G. Crosby, *J. Agric. Food Chem.* **22,** 842 (1974).
45. J. O. Nelson, P. C. Kearney, J. R. Plimmer, and R. E. Menzer, *Pestic. Biochem. Physiol.* **7,** 73 (1977).
46. G. P. Nilles and M. J. Zabik, *Pestic. Biochem. Physiol.* **22,** 684 (1974).
47. H. Igarashi, M. Uchiyama, and R. Sato *Noyaku Kagaku* **3,** 205 (1976).
48. N. B. K. Murthy and D. Kaufman, *J. Agric. Food Chem.* **26,** 1151 (1978).
49. M. Nakagawa and D. G. Crosby, *J. Agric. Food Chem.* **22,** 849 (1974).
50. J. P. Wargo, R. C. Honeycutt, and I. L. Alder, *J. Agric. Food Chem.* **23,** 1095 (1975).
51. L. M. Hunt, W. F. Chamberlain, B. N. Gilbert, D. E. Hopkins, and A. R. Gingrich, *J. Agric. Food Chem.* **25,** 1062 (1977).
52. Y. Niki, S. Kuwatsuka, and I. Yokomichi, *Agric. Biol. Chem.* **40,** 683 (1976).
53. R. K. Locke and R. L. Baron, *J. Agric. Food Chem.* **20,** 861 (1972).
54. I. L. Alder, B. M. Jones, and J. P. Wargo, Jr., *J. Agric. Food Chem.* **25,** 1339 (1977).
55. M. Nakagawa, K. Kawakubo, and M. Ishida, *Agric. Biol. Chem.* **35,** 764 (1971).
56. M. Nakagawa and M. Ando, *Agric. Biol. Chem.* **41,** 1975 (1977).
57. D. D. Focht and M. Alexander, *J. Agric. Food Chem.* **19,** 20 (1971).
58. R. V. Subba-Rao and M. Alexander, *J. Agric. Food Chem.* **25,** 855 (1977).
59a. C. M. Menzie, "Metabolism of Pesticides," p. 128. Bur. Sport Fish. & Wildlife Spec. Sci. Rep. Wildlife No. 127. Washington, D.C., 1969.
59b. I. P. Kapoor, R. L. Metcalf, R. F. Nystrom, and G. K. Sanga, *J. Agric. Food Chem.* **18,** 1145 (1970).
60. I. P. Kapoor, R. L. Metcalf, A. S. Hirwe, P.-Y. Lu, J. R. Coats, and R. F. Nystrom, *J. Agric. Food Chem.* **20,** 1 (1972).
61. S. Miyazaki and A. J. Throsteinson, *Bull. Environ. Contam. Toxicol.* **8,** 81 (1972).
62. K. C. Patil, F. Matsumura, and G. M. Boush, *Environ. Sci. Technol.* **6,** 629 (1972).
63. C. P. Rice and H. C . Sikka, *J. Agric. Food Chem.* **21,** 148 (1973).
64. G. Sundström, *J. Agric. Food Chem.* **25,** 18 (1977).
65. V. J. Feil, C. J. H. Lamoureux, E. Styrvoky, R. G. Zaylskie, E. J. Thacker, and G. M. Holman, *J. Agric. Food Chem.* **21,** 1072 (1973).
66. L. E. St. John, Jr. and D. J. Lisk, *J. Agric. Food Chem.* **21,** 644 (1973).
67. I. P. Kapoor, R. L. Metcalf, A. S. Hirwe, J. R. Coats, and M. S. Khalsa, *J. Agric. Food Chem.* **21,** 310 (1973).
68. A. S. Hirwe, R. L. Metcalf, and I. P. Kapoor, *J. Agric. Food Chem.* **20,** 818 (1972).
69. L. L. Miller, G. D. Nordblom, and G. A. Yost, *J. Agric. Food Chem.* **22,** 853 (1974).
70. N. Shindo, *Shokubutsu Boeki* **30,** 13 (1976).

71. R. C. Rhodes, *J. Agric. Food Chem.* **25,** 528 (1977).
72. W. D. Marshall and J. Singh, *J. Agric. Food Chem.* **25,** 1316 (1977).
73. J. P. Hubbell and J. E. Casida, *J. Agric. Food Chem.* **25,** 404 (1977)
74. K. Ishikawa, I. Okuda, and S. Kuwatsuka, *Agric. Biol. Chem.* **37,** 165 (1973).
75. K. Ishikawa, Y. Nakamura, and S. Kuwatsuka, *J. Pestic. Sci.* **1,** 49 (1976).
76. K. Ishikawa, Y. Nakamura, Y. Niki, and S. Kuwatsuka, *J. Pestic. Sci.* **2,** 17 (1977).
77. C. J. Soderquist, J. B. Bowers, and D. G. Crosby, *J. Agric. Food Chem.* **25,** 940 (1977).
78. J. R. DeBaun, D. L. Bova, C. K. Tseng, and J. J. Menn, *J. Agric. Food Chem.* **26,** 1098 (1978).
79. P. J. Murphy and T. L. Williams, *J. Med. Chem.* **15,** 137 (1972).
80. S. Kamimura, M. Nishikawa, H. Saeki, and Y. Takahi, *Phytopathology* **64,** 1273 (1974).
81. A. MacDonald, G. Chen, M. Kaykaty, and J. Fellig, *J. Agric. Food Chem.* **19,** 1222 (1971).
82. S. F. Krzeminski, C. K. Brackett, J. D. Fisher, and J. F. Spinnler, *J. Agric. Food Chem.* **23,** 1068 (1975).
83. S. Sumida, Y. Hisada, A. Kometani, and J. Miyamoto, *Agric. Biol. Chem.* **37,** 2127 (1973).
84. S. Sumida, R. Yoshihara, and J. Miyamoto, *J. Agric. Food Chem.* **37,** 2781 (1973).
85. G. W. Ivie, H. W. Dorough, and R. A. Cardona, *J. Agric. Food Chem.* **21,** 386 (1973).
86. Y. H. Atallah, D. M. Whitacre, and H. W. Dorough, *J. Agric. Food Chem.* **24,** 1007 (1976).
87. H. W. Dorough, D. M. Whitacre, and R. A. Cardona, *J. Agric. Food Chem.* **21,** 797 (1973).
88. K. Ishizuka, H. Hirata, and K. Fukunaga, *Agric. Biol. Chem.* **39,** 1431 (1975).
89. D. Ambrosi, P. C. Kearney, and J. A. Macchia, *J. Agric. Food Chem.* **25,** 868 (1977).
90. Y.-t. Woo, J. C. Arcos, M. F. Argus, G. W. Griffin, and K. Nishiyama *Biochem. Pharmacol.* **26,** 1535 (1977).
91. R. C. Rhodes, *J. Agric. Food Chem.* **25,** 1066 (1977).
92. R. H. Strang and R. L. Rogers, *J. Agric. Food Chem.* **22,** 1119 (1974).
93. A. Calderbank, Acta Phytopathol. Acad. *Sci. Hung.* **6,** 355 (1971).
94. H. Bratt, J. W. Daniel, and I. H. Monks, *Food Cosmet. Toxicol.* **10,** 489 (1972).
95. H. Ohkawa, Y. Hisada, N. Fujiwara, and J. Miyamoto, *Agric. Biol. Chem.* **38,** 1359 (1974).
96. N. Mikami, H. Yamamoto, Y. Tomida, and J. Miyamoto, *Jpn Pestic. Sci. Symp. 2nd,* p. 148 Abstr. (1977).
97. N. Mikami, H. Satogami, and J. Miyamoto, *J. Pestic. Sci.* **4,** 165 (1979).
98. N. L. Wolfe, R. G. Zepp, J. C. Doster, and R. C. Hollis, *J. Agric. Food Chem.* **24,** 1041 (1976).
99. K. Ogawa, H. Aizawa, and F. Yamauchi, *Adv. Pestic. Sci. Plenary Lect. Symp. Pap. Int. Congr. Pestic. Chem. 4th,* 1978, Main top. V-509.
100. R. W. Chadwick and J. J. Freal, *Bull. Environ. Contam. Toxicol.* **7,** 137 (1972).
100a. P. L. Grover and P. Sims, *Biochem. J.* **96,** 521 (1965).
101. I. Schuphan and K. Ballschmiter, *Nature* (*London*) **237,** 100 (1972).
102. N. M. Chopra and A. M. Mahfouz, *J. Agric. Food Chem.* **25,** 32 (1977).
103. S. Tashiro and F. Matsumura, *J. Agric. Food Chem.* **25,** 872 (1977).

104. V. J. Petrella, J. D. McKinney, J. P. Fox, and R. E. Webb, *J. Agric. Food Chem.* **25,** 393 (1977).
105. A. Richardson and J. Robinson, *Xenobiotica* **1,** 213 (1971).
106. K. C. Patil, F. Matsumura, and G. M. Boush, *Environ. Sci. Technol.* **6,** 629 (1972).
107. M. A. Q. Khan, A. Kamal, R. J. Wolin, and J. Runnels, *Bull. Environ. Contam. Toxicol.* **8,** 219 (1972).
108. C. H. Walker and G. A. El Zorgani, *Arch. Environ. Contam. Toxicol.* **2,** 97 (1974).
109. J. P. Lay, W. Klein, and F. Korte, *Chemosphere* **5,** 193 (1974).
110. C. J. Whitten and D. L. Bull, *J. Agric. Food Chem.* **22,** 234 (1974).
111. W.-T. Chin, W. C. Duane, M. B. Szalkowski, and D. E. Stallard, *J. Agric. Food Chem.* **23,** 963 (1975).
112. R. E. Holm, W.-T, Chin, D. H. Wagner, and D. E. Stallard, *J. Agric. Food Chem.* **23,** 1056 (1975).
113. W.-T. Chin, W. C. Duane, D. L. Ballee, and D. E. Stallard, *J. Agric. Food Chem.* **24,** 1071 (1976).
114. W. J. Bartley, N. R. Andrawes, E. L. Chancy, W. P. Bagley, and H. W. Spurr, *J. Agric. Food Chem.* **18,** 446 (1970).
115. N. R. Andrawes, W. P. Bagley, and R. A. Herrett, *J. Agric. Food Chem.* **19,** 727 (1971).
116. F. A. Richey, Jr., W. J. Bartley, and K. P. Sheets, *J. Agric. Food Chem.* **25,** 47 (1977).
117. B. W. Hichs, H. W. Dorough, and H. M. Mehendale, *J. Agric. Food Chem.* **20,** 151 (1972).
118. J. Harvey, Jr. and J.C.-Y. Han, *J. Agric. Food Chem.* **26,** 902 (1978).
119. J. E. Hill and R. I. Krieger, *J. Agric. Food Chem.* **23,** 1125 (1975).
120. D. G. Crosby and A. S. Wong, *J. Agric. Food Chem.* **21,** 1049 (1973).
121. C.-S. Feung, R. H. Hamilton, and R. O. Mumma, *J. Agric. Food Chem.* **23,** 373 (1975).
122. C.-S. Feung, R. H. Hamilton, and R. O. Mumma, *J. Agric. Food Chem.* **24,** 1013 (1976).
123. C.-S. Feung, S. L. Loerch, R. H. Hamilton, and R. O. Mumma, *J. Agric. Food Chem.* **26,** 1064 (1978).
124. S. Wathana and F. T. Corbin, *J. Agric. Food Chem.* **20,** 23 (1972).
125. D. G. Crosby and A. S. Wong, *J. Agric. Food Chem.* **21,** 1052 (1973).
126. S. G. Gorbach, K. Kuenzler, and J. Asshauer, *J. Agric. Food Chem.* **25,** 507 (1977).
127. A. E. Smith, *J. Agric. Food Chem.* **25,** 893 (1977).
128. T. Tokuda, M. Nishiki, G. Hoshi, K. Shinoda, M. Ishida, and T. Misato, *J. Pestic. Sci.* **1,** 283 (1976).
129. Y. Soeda, S. Kosaka, and T. Noguchi, *Agric. Biol. Chem.* **36,** 817 (1972).
130. Y. Soeda, S. Kosaka, and T. Noguchi, *Agric. Biol. Chem.* **36,** 931 (1972).
131. Y. Yasuda, S. Hashimoto, Y. Soeda, and T. Noguchi *Ann. Phytopathol. Sci. Jpn.* **39,** 49 (1973).
132. J. R. Fleeker, H. M. Lacy, I. R. Schultz, and E. C. Houkom, *J. Agric. Food Chem.* **22,** 592 (1974).
133. P. G. C. Douch, *Xenobiotica* **4,** 457 (1974).
134. J. J. Kirkland, *J. Agric. Food Chem.* **21,** 171 (1973).
135. J. A. Gardiner, J. J. Kirkland, H. L. Klopping, and H. Sherman, *J. Agric. Food Chem.* **22,** 419 (1974).
136. P. G. C. Douch, *Xenobiotica* **3,** 367 (1973).
137. J. R. Fleeker and H. M. Lacy, *J. Agric. Food Chem.* **25,** 51 (1977).

138. J. P. Rouchaud, J. R. Decallonne, and J. A. Meyer, *Phytopathology* **64,** 1513 (1974).
139. J. P. Rouchaud, J. R. Decallonne, and J. A. Meyer *Pestic. Sci.* **8,** 31 (1977).
140. C. J. DiCuollo, J. A. Miller, W. L. Mendelson, and J. F. Pagano, *J. Agric. Food Chem.* **22,** 948 (1974).
141. W. Mittelstaedt, G. G. Still, H. Dürbeck, and F. Führ, *J. Agric. Food Chem.* **25,** 908 (1977).
142. M. Uchiyama, H. Abe, R. Sato, M. Shimura, and T. Watanabe, *Agric. Biol. Chem.* **37,** 737 (1973).
143. H. Ohkawa, R. Yoshihara, T. Kohara, and J. Miyamoto, *Agric. Boil. Chem.* **38,** 1035 (1974).
144. K. Ogawa, M. Tsuda, F. Yamauchi, I. Yamaguchi, and T. Misato, *J. Pestic. Sci.* **2,** 51 (1977).
145. K. Ogawa, M. Tsuda, F. Yamauchi, I. Yamaguchi, and T. Misato, *J. Pestic. Sci.* **1,** 219 (1976).
146. T. Suzuki and M. Takeda, *Chem. Pharm. Bull.* **24,** 1967 (1976).
147. T. Suzuki and M. Takeda, *Chem. Pharm. Bull.* **24,** 1983 (1976).
148. T. Suzuki and M. Takeda, *Chem. Pharm. Bull.* **24,** 1988 (1976).
149. P. G. C. Douch and J. N. Smith, *Biochem. J.* **125,** 385 (1971).
150. B. V. Tucker and D. E. Pack, *J. Agric. Food Chem.* **20,** 412 (1972).
151. M. Slade and J. E. Casida, *J. Agric. Food Chem.* **18,** 467 (1970).
152. R. K. Locke, V. B. Bastone, and R. L. Baron, *J. Agric. Food Chem.* **19,** 1205 (1971).
153. H. W. Dorough, *J. Agric. Food Chem.* **18,** 1015 (1970).
154. H.-M. Cheng and J. E. Casida, *J. Agric. Food Chem.* **21,** 1037 (1973).
155. R. J. Kuhr, *J. Agric. Food Chem.* **18,** 1023 (1970).
156. S. P. Shrivastava, G. P. Georghiou, and T. R. Fukuto, *Entomol. Exp. Appl.* **14,** 333 (1971).
157. H. J. Benezet and F. Matsumura, *J. Agric. Food Chem.* **22,** 427 (1974).
158. R. W. Meikle, *Bull. Environ. Contam. Toxicol.* **10,** 29 (1973).
159. B. R. Sonawane and C. O. Knowles, *Pestic. Biochem. Physiol.* **1,** 472 (1972).
160. S.-Y. Liu and J. M. Bollag, *J. Agric. Food Chem.* **19,** 487 (1971).
161. J. M. Bollag and S.-Y. Liu, *Nature* (*London*) **236,** 177 (1972).
162. T. H. Lin, H. H. North, and R. E. Menzer, *J. Agric. Food Chem.* **23,** 253 (1975).
163. G. N. Guirguis and W. A. Brindley, *J. Agric. Food Chem.* **23,** 274 (1975).
164. J. B. Knaak, D. M. Munger, and J. McCarthy, *J. Agric. Food Chem.* **18,** 827 (1970).
165. R. J. Ashworth and T. J. Sheets, *J. Agric. Food Chem.* **20,** 407 (1972).
166. D. J. Pree and J. L. Saunders, *J. Agric. Food Chem.* **22,** 620 (1974).
167. T. C. Marshall and H. W. Dorough, *J. Agric. Food Chem.* **25,** 1003 (1977).
168. L. S. Jordan, A. A. Jurqiyah, A. R. De Mur, and W. A. Clerx, *J. Agric. Food Chem.* **23,** 286 (1975).
169. D. S. Frear and H. R. Swanson, *Phytochemistry* **11,** 1919 (1972).
170. F. S. Tanaka, R. G. Wien, and R. G. Zaylskie, *J. Agric. Food Chem.* **25,** 1068 (1977).
171. I. Schuphan, *Chemosphere* **3,** 131 (1974).
172. A. Haque, I. Weisgerber, D. Kotzias, and W. Klein, *Pestic. Biochem. Physiol.* **7,** 321 (1977).
173. I. Takase, T. Nakahara, and K. Ishizuka, *J. Pestic. Sci.* **3,** 9 (1978).
174. R. L. Metcalf, P.-Y. Lu, and S. Bowlus, *J. Agric. Food Chem.* **23,** 359 (1975).
175. K. T. Koshy, A. R. Friedman, A. L. VanDerSlik, and D. R. Graber, *J. Agric. Food Chem.* **23,** 1084 (1975).
176. C. S. Crecelius and C. O. Knowles, *J. Agric. Food Chem.* **26,** 486 (1978).

177. D. M. Whitacre, M. Badie, B. A. Schwemmer, and L. I. Diaz, *Bull. Environ. Contam. Toxicol.* **16,** 689 (1976).
178. K. Endo, Y. Mori, K. Kakiki, and T. Misato, *Nippon Nogei Kagaku Kaishi* **44,** 356 (1970).
179. Y. Uesugi and C. Tomizawa, *Agric. Biol. Chem.* **36,** 313 (1971).
180. J. B. McBain, L. J. Hoffman, and J. J. Menn, *J. Agric. Food Chem.* **18,** 1139 (1970).
181. J. B. McBain, L. J. Hoffman, and J. J. Menn, *Pestic. Biochem. Physiol.* **1,** 356 (1971).
182. S. J. Flashinski and E. P. Lichtenstein, *Can. J. Microbiol.* **20,** 399 (1974).
183. E. P. Lichtenstein, H. Parlar, F. Korte, and A. Suss, *J. Agric. Food Chem.* **25,** 845 (1977).
184. J. B. Miaullis, L. J. Hoffman, J. R. DeBaun, and J. J. Menn, *J. Agric. Food Chem.* **25,** 501 (1977).
185. R. J. Bussey, M. A. Christenson, and M. S. O'Connor, *J. Agric. Food Chem.* **25,** 993 (1977).
186. D. L. Bull and R. L. Ridgway, *J. Agric. Food Chem.* **17,** 837 (1969).
187. M. Chiba, S. Kato, and I. Yamamoto, *J. Pestic. Sci.* **1,** 179 (1976).
188. M. D. Gilbert, S. P. Monselise, L. J. Edgerton, G. A. Maylin, L. J. Hicks, and D. J. Lisk, *J. Agric. Food Chem.* **23,** 290 (1975).
189. M. L. Rueppel, B. B. Brightwell, J. Schaefer, and J. T. Marvel, *J. Agric. Food Chem.* **25,** 517 (1977).
190. K. Mihara, K. Nambu, Y. Misaki, and J. Miyamoto, *J. Pestic. Sci.* **1,** 207 (1976).
191. H. Ohkawa, N. Mikami, and J. Miyamoto, *Agric. Biol. Chem.* **40,** 2125 (1976).
192. N. Mikami, H. Ohkawa, and J. Miyamoto, *J. Pestic. Sci.* **2,** 119 (1977).
193. J. S. Thornton, *J. Agric. Food Chem.* **19,** 890 (1971).
194. T. B. Waggoner, *J. Agric. Food Chem.* **20,** 157 (1972).
195. J. E. Bakke and C. E. Price, *J. Environ. Sci. Health, Part B* **12,** 251 (1977).
196. J. C. Pekas, J. E. Bakke, J. L. Giles, and C. E. Price, *J. Environ. Sci. Health Part B* **12,** 261 (1977).
197. T.-S. Kao and T. R. Fukuto, *Pestic. Biochem. Physiol.* **7,** 83 (1977).
198. T. R. Fukuto, S. P. Shrivastava, and A. L. Black, *Pestic. Biochem. Physiol.* **2,** 162 (1972).
199. J. E. Bakke, V. J. Feil, C. E. Fjelstul, and E. J. Thacker, *J. Agric. Food Chem.* **20,** 384 (1972).
200. I. P. Kapoor and R. C. Blinn, *J. Agric. Food Chem.* **25,** 413 (1977).
201. J. Zulalian and R. C. Blinn, *J. Agric. Food Chem.* **25,** 1033 (1977).
202. L. E. Wendel and D. L. Bull, *J. Agric. Food Chem.* **18,** 420 (1970).
203. D. H. Hutson, E. C. Hoadley, and B. A. Pickering, *Xenobiotica* **1,** 593 (1971).
204. D. H. Hutson and E. C. Hoadley, *Xenobiotica* **2,** 107 (1972).
205. M. J. Crawford, D. H. Hutson, and P. A. King, *Xenobiotica* **6,** 745 (1976).
206. M. H. Akhtar and T. S. Foster, *J. Agric. Food Chem.* **25,** 1017 (1977).
207. H. Geissbühler, G. Voss, and R. Anliker, *Residue Rev.* **37,** 39 (1971).
208. G. W. Lucier and R. E. Menzer, *J. Agric. Food Chem.* **19,** 1249 (1971).
209. C. Tomizawa and Y. Uesugi, *Agric. Biol. Chem.* **36,** 294 (1972).
210. H. Yamamoto, C. Tomizawa, Y. Uesugi, and T. Murai, *Agric. Biol. Chem.* **37,** 1553 (1973).
211. W. F. Chamberlain and D. E. Hopkins, *J. Econ. Entomol.* **66,** 119 (1973).
212. E. P. Lichtenstein, T. W. Fuhremann, A. A. Hochberg, R. N. Zahlten, and F. W. Stratman, *J. Agric. Food Chem.* **21,** 416 (1973).
213. T. Suzuki and M. Uchiyama, *J. Agric. Food Chem.* **23,** 281 (1975).

214. J. Miyamoto, N. Mikami, K. Mihara, Y. Takimoto, H. Kohda, and H. Suzuki, *J. Pestic. Sci.* **3,** 35 (1978).
215. D. G. Rowlands and C. E. Dyte, *Pestic. Sci.* **3,** 191 (1972).
216. R. Greenhalgh and W. D. Marshall, *J. Agric. Food Chem.* **24,** 708 (1976).
217. A. Wakimura and J. Miyamoto, *Agric. Biol. Chem.* **35,** 410 (1971).
218. N. Mikami, H. Ohkawa, and J. Miyamoto, *J. Pestic. Sci.* **1,** 273 (1976).
219. M. Stiasni, W. Deckers, K. Schmidt, and H. Simon, *J. Agric. Food Chem.* **17,** 1017 (1969).
220. M. C. Bowman and K. R. Hill, *J. Agric. Food Chem.* **19,** 342 (1971).
221. J. G. Leesch and T. R. Fukuto, *Pestic. Biochem. Physiol.* **2,** 223 (1972).
222. M. Ando, Y. Iwasaki, and M. Nakagawa, *Agric. Biol. Chem.* **39,** 2137 (1975).
223. C. Hutacharern and C. O. Knowles, *Bull. Environ. Contam. Toxicol.* **13,** 351 (1975).
224. A. F. Machin, M. P. Quick, and N. F. Janes, *Chem. Ind.* (*London*) **16,** 1198 (1971).
225. R. S. H. Yang, E. Hodgson, and W. C. Dauterman, *J. Agric. Food Chem.* **19,** 10 (1971).
226. R. S. H. Yang, E. Hodgson, and W. C. Dauterman, *J. Agric. Food Chem.* **19,** 14 (1971).
227. N. F. Janes, A. F. Machin, M. P. Quick, H. Rogers, D. E. Mundy, and A. J. Cross, *J. Agric. Food Chem.* **21,** 121 (1973).
228. A. F. Machin, M. P. Quick, H. Rogers, and N. F. Janes, *Bull. Environ. Contam. Toxicol.* **7,** 270 (1972).
229. A. F. Machin, *J. Pestic. Sci.* **4,** 425 (1973).
230. M. C. Bowman and D. B. Leuck, *J. Agric. Food Chem.* **19,** 1215 (1971).
231. W. A. Mason and C. E. Meloan, *J. Agric. Food Chem.* **24,** 299 (1976).
232. K. Mihara and J. Miyamoto, *Agric. Biol. Chem.* **38,** 1913 (1974).
233. Y. Uesugi and C. Tomizawa, *Agric. Biol. Chem.* **35,** 941 (1971).
234. I. Ueyama, Y. Uesugi, C. Tomizawa, and T. Murai, *Agric. Biol. Chem.* **37,** 1543 (1973).
235. I. Takase, K. E. Tan, and K. Ishizuka, *Agric. Biol. Chem.* **37,** 1563 (1973).
236. R. E. Menzer, Z. M. Iqbal, and G. R. Boyd, *J. Agric. Food Chem.* **19,** 351 (1971).
237. Z. M. Iqbal and R. E. Menzer, *Biochem. Pharmacol.* **21,** 1569 (1972).
238. S. M. A. D. Zayed, I. M. I. Fakhr, and M. R. E. Bahig, *Biochem. Pharmacol.* **22,** 285 (1973).
239. G. W. Lucier and R. E. Menzer, *J. Agric. Food Chem.* **18,** 698 (1970).
240. H. H. Sauer, *J. Agric. Chem.* **20,** 578 (1972).
241. D. L. Bull, C. J. Whitten, and G. W. Ivie, *J. Agric. Chem.* **24,** 601 (1976).
242. J. E. Cassidy, D. P. Ryskiewich, and R. T. Murphy, *J. Agric. Chem.* **17,** 558 (1969).
243. D. O. Eberle and W. D. Hörmann, *J. Assoc. Off. Anal. Chem* **54,** 150 (1971).
244. D. Ambrosi, P. C. Kearney, and J. A. Macchia, *J. Agric. Food Chem.* **25,** 342 (1977).
245. W. H. Harned and J. E. Casida, *J. Agric. Food Chem.* **24,** 689 (1976).
246. M. Elliott, N. F. Janes, E. C. Kimmel, and J. E. Casida, *J. Agric. Food Chem.* **20,** 300 (1972).
247. J. Miyamoto, T. Nishida, and K. Ueda, *Pestic. Biochem. Physiol.* **1,** 293 (1971).
248. K. Ueda, L. C. Gaughan, and J. E. Casida, *J. Agric. Food Chem.* **23,** 106 (1975).
249. T. R. Roberts and M. E. Standen, *Pestic. Sci.* **8,** 305 (1977).
250. L. O. Ruzo, R. L. Holmstead, and J. E. Casida, *J. Agric. Food Chem.* **25,** 1385 (1977).

251. L. C. Gaughan, T. Unai, and J. E. Casida, *J. Agric. Food Chem.* **25,** 9 (1977).
252. H. Ohkawa, H. Kaneko, H. Tsuji, and J. Miyamoto, *J. Pestic. Sci.* **4,** 143 (1979).
253. J. C. Ramsey, J. Q. Rose, W. H. Braun, and P. J. Gehring, *J. Agric. Food Chem.* **22,** 870 (1974).
254. A. J. Pik, E. Peake, M. T. Strosher, and G. W. Hodgson, *J. Agric. Food Chem.* **25,** 1054 (1977).
255. C.-C. Yu, G. M. Booth, D. J. Hansen, and J. R. Larsen, *J. Agric. Food Chem.* **23,** 309 (1975).
256. G. L. Lamoureux, R. H. Shimabukuro, H. R. Swanson, and D. S. Frear, *J. Agric. Food Chem.* **18,** 81 (1970).
257. G. L. Lamoureux, L. E. Stafford, and R. H. Shimabukuro, *J. Agric. Food Chem.* **20,** 1004 (1972).
258. G. L. Lamoureux, L. E. Stafford, R. H. Shimabukuro, and R. G. Zaylskie, *J. Agric. Food Chem.* **21,** 1020 (1973).
259. R. H. Shimabukuro, W. C. Walsh, G. L. Lamoureux, and L. E. Stafford, *J. Agric. Food Chem.* **21,** 1031 (1973).
260. S. U. Khan and T. S. Foster, *J. Agric. Food Chem.* **24,** 768 (1976).
261. C. G. P. Pillai, J. D. Weete, and D. E. Davis, *J. Agric. Food Chem.* **25,** 852 (1977).
262. S. U. Khan and P. B. Marriage, *J. Agric. Food Chem.* **25,** 1408 (1977).
263. J. V. Crayford and D. H. Hutson, *Pestic. Biochem. Physiol.* **2,** 295 (1972).
264. S. C. Lau, D. B. Katague, and D. W. Stoutamire, *J. Agric. Food Chem.* **21,** 1091 (1973).
265. J. E. Bakke and C. E. Price, *J. Agric. Food Chem.* **21,** 640 (1973).
266. G. L. Larsen and J. E. Bakke, *J. Agric. Food Chem.* **23,** 388 (1975).
267. J. E. Bakke, J. D. Robbins, and V. J. Feil, *J. Agric. Food Chem.* **19,** 462 (1971).
268. J. S. Thornton and C. W. Stanley, *J. Agric. Food Chem.* **25,** 380 (1977).
269. M. J. Lynch and S. K. Figdor, *J. Agric. Food Chem.* **25,** 1344 (1977).
270. S. Yllner, *Acta Pharmacol. Toxicol.* **30,** 69 (1971).
271. R. S. Horvath, *J. Agric. Food Chem.* **19,** 291 (1971).
272. H. C. Sikka, R. S. Lynch, and M. Lindenberger, *J. Agric. Food Chem.* **22,** 230 (1974).
273. S. Miyazaki, H. C. Sikka, and R. S. Lynch, *J. Agric. Food Chem.* **23,** 365 (1975).
274. G. W. Ivie, D. E. Clark, and D. D. Rushing, *J. Agric. Food Chem.* **22,** 632 (1974).
275. R. C. Rhodes, H. L. Pease, and R. K. Brantley, *J. Agric. Food Chem.* **19,** 745 (1971).
276. J. L. Buckland, R. F. Collins, and E. M. Pullin, *J. Pestic. Sci.* **4,** 149 (1973).
277. Y. Fujii, T. Kurokawa, S. Ishida, I. Yamaguchi, and T. Misato, *J. Pestic. Sci.* **1,** 313 (1976).
278. W. W. Shindy, L. S. Jordan, V. A. Jolliffe, C. W. Coggins, Jr., and J. Kumamoto, *J. Agric. Food Chem.* **21,** 629 (1973).

Author Index

Numbers in parentheses are reference numbers and indicate that an author's work is referred to although the name is not cited in the text. Numbers in italics show the page on which the complete reference is listed.

A

Abe, H., 103 (142), *216*
Aizawa, H., 74 (99), *214*
Akhtar, M. H., 155 (206), *217*
Adler, I. L., 33, 35 (50, 54), *213*
Allen, J. L., 25 (38), *213*
Alexander, M., 38 (57, 58), 42 (57, 58), *213*
Ambrosi, D., 65 (89), 180 (244), *214, 218*
Anagstopoulos, E., 20 (32), *212*
Ando, M., 36 (56), 167 (222), *213, 218*
Andrawes, N. R., 86 (114, 115), *215*
Anliker, R., 156 (207), *217*
Arcos, J. C., 66 (90), *214*
Argus, M. F., 66 (90), *214*
Ashworth, R. J., 120 (165), *216*
Asshauer, J., 95 (126), *215*
Atallah, Y. H., 64 (86), *214*

B

Badie, M., 132 (177), *217*
Bagley, W. P., 86 (114, 115), *215*
Bahig, M. R. E., 176 (238), *218*
Bakke, J. E., 146 (195, 196), 149 (199), 198 (265, 266), 199 (267), *217, 219*
Ballee, D. L., 85 (113), *215*
Ballschmiter, K., 77 (101), *214*
Bandal, S. K., 26 (39), *213*
Baron, R. L., 34 (53), 112 (152), 116 (152), 119 (152), *213, 216*
Bartley, W. J., 86 (114, 116), *215*
Bastone, V. B., 112 (152), 116 (152), 119 (152), *216*
Benezet, H. J., 17 (27), 111 (157), 117 (157), *212, 216*
Berggren, M., 22 (34), *212*
Bingham, S. W., 8 (9, 12), *212*
Black, A. L., 148 (198), *217*
Blinn, R. C., 18 (29), 150 (200), 151 (201), *212, 217*
Bollag, J. M., 20 (31), 119 (160, 161), *212, 216*
Boush, G. M., 39 (62), 81 (106), *213, 215*
Bova, D. L., 55 (78), *214*
Bowers, J. B., 57 (77), *214*
Bowlus, S., 128 (174), *216*
Bowman, M. C., 150 (220), 165 (220), 170 (230), *218*
Boyd, G. R., 174 (236), *218*
Brackett, C. K., 62 (82), *214*
Brantley, R. K., 207 (275), *219*
Bratt, H., 69 (94), *214*
Braun, W. H., 191 (253), *219*
Briggs, G. G., 23 (35), *212*
Brightwell, B. B., 140 (189), *217*
Brindley, W. A., 119 (163), *216*
Buckland, J. L., 208 (276), *219*
Bull, D. L., 85 (110), 138 (186), 153 (202), 178 (241), *215, 217, 218*
Bussey, R. J., 136 (185), *217*

C

Calderbank, A., 69 (93), *214*
Cardona, R. A., 64 (85, 87), *214*
Casida, J. E., 26 (39), 55 (73), 111 (151), 113 (154), 181 (245), 183 (246), 184 (246), 185 (248), 187 (250), 188 (251), *213, 214, 216, 218, 219*
Cassidy, J. E., 179 (242), *218*
Chadwick, R. W., 76 (100), *214*
Chamberlain, W. F., 33, 159 (51, 211), *213, 217*
Champagne, D. A., 18 (29), *212*
Chancy, E. L., 86 (114), *215*
Chang, K. M., 17 (26), *212*
Chen, C.-C., 10 (15), *212*

Chen, G., 61 (81), *214*
Chen, Y.-L., 10 (15), *212*
Cheng, H.-M., 113 (154), *216*
Chiba, M., 139 (187), 163 (187), *217*
Chin, W.-T., 13 (18), 85 (111, 112, 113), *212, 215*
Chopra, N. M., 78 (102), *214*
Christenson, M. A., 136 (185), *217*
Clark, D. E., 206 (274), *219*
Clerx, W. A., 123 (168), *216*
Coats, J. R., 39 (60), 44 (67), 45 (60), 48 (67), 50 (60), 51 (67), *213*
Coggins, C. W., Jr., 211 (278), *219*
Collins, R. F., 208 (276), *219*
Corbin, F. T., 92 (124), *215*
Cox, B. L., 8 (11), *212*
Crawford, M. J., 155 (205), *217*
Crayford, J. V., 197 (263), *219*
Crecelius, C. S., 130 (176), *216*
Crosby, D. G., 29 (43, 44), 33 (49), 57 (77), 90 (120), 94 (125), *213, 214, 215*
Cross, A. J., 169 (227), *218*

D

Daniel, J. W., 69 (94), *214*
Darda, S., 14 (20, 21), *212*
Darskus, R. L., 14 (20), *212*
Dauterman, W. C., 169 (225, 226), *218*
Davis, D. E., 196 (261), *219*
DeBaun, J. R., 55 (78), 136 (184), *214, 217*
Decallonne, J. R., 14 (19), 100 (138, 139), *212, 216*
Deckers, W., 164 (219), *218*
De Mur, A. R., 123 (168), *216*
Diaz, L. I., 132 (177), *217*
Di Cuollo, C. J., 101 (140), *216*
Dorough, H. W., 64 (85, 86, 87), 86 (117), 112 (153), 115 (153), 119 (153), 120 (153, 167), 121 (153), *214, 215, 216*
Doster, J. C., 73 (98), *214*
Douch, P. G. C., 98 (133), 99 (136), 100 (136), 108 (149), *215, 216*
Duane, W. C., 85 (111, 113), *215*
Dürbeck, H., 102 (141), *216*
Dyte, C. E., 162 (215), *218*

E

Eberle, D. O., 179 (243), *218*
Edgerton, L. J., 140 (188), *217*
Eichler, D., 14 (20), *212*
Elliott, M., 183 (246), 184 (246), *218*
El Zorgani, G. A., 82 (108), *215*
Endo, K., 133 (178), *217*

F

Fakhr, I. M. I., 176 (238), *218*
Feil, V. J., 21 (33), 29 (42), 41 (65), 149 (199), 199 (267), *212, 213, 217, 219*
Fellig, J., 61 (81), *214*
Feung, C.-S., 91 (121, 122, 123), *215*
Figdor, S. K., 201 (269), *219*
Fisher, J. D., 3 (3), 62 (82), *212, 214*
Fjelstul, C. E., 149 (199), *217*
Flashinski, S. J., 135 (182), *217*
Fleeker, J. R., 98 (132), 100 (137), *215*
Focht, D. D., 38 (57), 42 (57), *213*
Foster, T. S., 155 (206), 196 (260), *217, 219*
Fox, J. P., 80 (104), *215*
Freal, J. J., 76 (100), *214*
Frear, D. S., 124 (169), 195 (256), *216, 219*
Friedman, A. R., 129 (175), *216*
Führ, F., 102 (141), *216*
Fuhremann, T. W., 161 (212), *217*
Fujii, Y., 210 (277), *219*
Fujiwara, N., 71 (95), *214*
Fukunaga, K., 65 (88), *214*
Fukuto, T. R., 114 (156), 147 (197), 148 (198), 166 (221), *216, 217, 218*

G

Gardiner, J. A., 99 (135), *215*
Gatterdam, P. E., 18 (28), *212*
Gaughan, L. C., 185 (248), 188 (251), *218, 219*
Gehring, P. J., 191 (253), *219*
Geissbühler, H., 156 (207), *217*
Georghiou, G. P., 114 (156), *216*
Gilbert, B. N., 33 (51), *213*
Gilbert, M. D., 140 (188), *217*
Giles, J. L., 146 (196), *217*
Gingrich, A. R., 33 (51), *213*
Gorbach, S. G., 95 (126), *215*
Graber, D. R., 129 (175), *216*
Greenhalgh, R., 162 (216), *218*
Griffin, G. W., 66 (90), *214*
Grover, P. L., 76 (100a), *214*
Guirguis, G. N., 119 (163), *216*

H

Hagedorn, M. L., 10 (14), *212*
Hamilton, R. H., 91 (121, 122, 123), *215*
Han, J. C.-Y., 87 (118), *215*
Hansen, D. J., 193 (255), *219*
Haque, A., 126 (172), *216*
Harned, W. H., 181 (245), *218*

Harvey, J., Jr., 87 (118), *215*
Hashimoto, S., 98 (131), *215*
Herrett, R. A., 86 (115), *215*
Hichs, B. W., 86 (117), *215*
Hicks, L. J., 140 (188), *217*
Hill, J. E., 88 (119), *215*
Hill, K. R., 150 (220), 165 (220), *218*
Hirata, H., 65 (88), *214*
Hirwe, A. S., 39 (60), 44 (67), 45 (60), 48 (67, 68), 50 (60), 51 (67, 68), *213*
Hisada, Y., 63 (83), 71 (95), *214*
Hoadley, E. C., 154 (203, 204), *217*
Hochberg, A. A., 161 (212), *217*
Hodgson, E., 169 (225, 226), *218*
Hodgson, G. W., 192 (254), *219*
Hörmann, W. D., 179 (243), *218*
Hoffman, L. J., 134 (180, 181), 136 (184), *217*
Hogan, J. W., 25 (38), *213*
Hollis, R. C., 73 (98), *214*
Holm, R. E., 85 (112), *215*
Holman, G. M., 41 (65), *213*
Holmstead, R. L., 187 (250), *218*
Honeycutt, R. C., 33 (50), *213*
Hopkins, D. E., 33 (51), 159 (211), *213, 217*
Hornish, R. E., 12 (17), *212*
Horvath, R. S., 204 (271), *219*
Hoshi, G., 97 (128), *215*
Houkom, E. C., 98 (132), *215*
Hubbell, J. P., 55 (73), *214*
Hunt, L. M., 33 (51), *213*
Hutacharern, C., 168 (223), *218*
Hutson, D. H., 154 (203, 204), 155 (205), 197 (263), *217, 219*
Hutzinger, O., 2 (1), 11 (16), *212*

I

Igarashi, H., 31 (47), *213*
Iqbal, Z. M., 174 (236, 237), 177 (237), *218*
Isenssee, A. R., 27 (41), *213*
Ishida, M., 36 (55), 97 (128), *213, 215*
Ishida, S., 210 (277), *219*
Ishikawa, K., 56 (74, 75, 76), *214*
Ishizuka, K., 65 (88), 127 (173), 173 (235), *214, 216, 218*
Ivie, G. W., 15 (22), 64 (85), 178 (241), 206 (274), *212, 214, 218, 219*
Iwasaki, Y., 167 (222), *218*

J

Jacobsen, A. M., 15 (23), 21 (33), *212*
Janes, N. F., 169 (224, 227, 228), 183 (246), 184 (246), *218*
Jolliffe, V. A., 211 (278), *219*
Jones, B. M., 35 (54), *213*
Jordan, L. S., 123 (168), 211 (278), *216, 219*
Jurqiyah, A. A., 123 (168), *216*

K

Kakiki, K., 133 (178), *217*
Kamal, A., 81 (107), *215*
Kamimura, S., 60 (80), *214*
Kaneko, H., 189 (252), *219*
Kao, T.-S., 147 (197), *217*
Kapoor, I. P., 39 (59b, 60), 44 (67), 45 (60), 48 (67, 68), 50 (60), 51 (67, 68), 150 (200), *213, 217*
Katague, D. B., 197 (264), *219*
Kato, S., 139 (187), 163 (187), *217*
Kaufman, D., 31 (48), *213*
Kaufman, D. D., 20 (30), *212*
Kawakubo, K., 36 (55), *213*
Kaykaty, M., 61 (81), *214*
Kearney, P. C., 27 (41), 28 (45), 29 (45), 65 (89), 180 (244), *213, 214, 218*
Khalsa, M. S., 44 (67), 48 (67), 51 (67), *213*
Khan, M. A. Q., 81 (107), *215*
Khan, S. U., 196 (260, 262), *219*
Kimmel, E. C., 183 (246), 184 (246), *218*
King, P. A., 155 (205), *217*
Kirkland, J. J., 99 (134, 135), *215*
Klein, W., 20 (32), 83 (109), 126 (172), *212, 215, 216*
Klingebiel, U. I., 20 (30), 27 (40, 41), *212, 213*
Klopping, H. L., 99 (135), *215*
Knaak, J. B., 120 (164), *216*
Knowles, C. O., 17 (25, 26, 27), 118 (159), 130 (176), 168 (223), *212, 216, 218*
Kohara, T., 105 (143), *216*
Kohda, H., 162 (214), *218*
Kometani, A., 63 (83), *214*
Königer, M., 11 (16), *212*
Korte, F., 20 (32), 83 (109), 135 (183), *212, 215, 217*
Kosaka, S., 98 (129, 130), *215*
Koshy, K. T., 129 (175), *216*
Kosuge, T., 24 (36), *212*
Kotzias, D., 126 (172), *216*
Krieger, R. I., 88 (119), *215*
Krzeminski, L. F., 8 (11), *212*
Krzeminski, S. F., 62 (82), *214*
Kuenzler, K., 95 (126), *215*
Kuhr, R. J., 114 (155), 119 (155), *216*
Kumamoto, J., 211 (278), *219*
Kurokawa, T., 210 (277), *219*

Kuwatsuka, S., 4 (5), 34 (52), 56 (5, 74, 75, 76), 93 (5), 209 (5), *213, 214*

L

Laanio, T. L., 27 (41), *213*
Lacy, H. M., 98 (132), 100 (137), *215*
Lamoureux, C. J. H., 41 (65), *213*
Lamoureux, G. L., 7 (8), 15 (24), 195 (256, 257, 258), 196 (257, 259), *212, 219*
Larsen, G. L., 198 (266), *219*
Larsen, J. R., 193 (255), *219*
Lau, S. C., 197 (264), *219*
Lay, J. P., 83 (109), *215*
Leesch, J. G., 166 (221), *218*
Leitis, E., 29 (44), *213*
Leuck, D. B., 170 (230), *218*
Lichtenstein, E. P., 135 (182, 183), 161 (212), *217*
Lin, T. H., 119 (162), *216*
Lindenberger, M., 205 (272), *219*
Lisk, D. J., 43 (66), 140 (188), *213, 217*
Liu, S.-Y., 119 (160, 161), *216*
Locke, R. K., 34 (53), 112 (152), 116 (152), 119 (152), *213, 216*
Loerch, S. L., 91 (123), *215*
Lu, P.-Y., 39 (60), 45 (60), 50 (60), 128 (174), *213, 216*
Lucier, G. W., 156 (208), 177 (239), *217, 218*
Lynch, M. J., 201 (269), *219*
Lynch, R. S., 205 (272, 273), *219*

M

McBain, J. B., 134 (180, 181), *217*
McCarthy, J., 120 (164), *216*
Macchia, J. A., 65 (89), 180 (244), *214, 218*
MacDonald, A., 61 (81), *214*
McGahen, L. L., 9 (13), *212*
Machin, A. F., 169, (224, 227, 228, 229), *218*
McKinney, J. D., 80 (104), *215*
Mahfouz, A. M., 78 (102), *214*
Marriage, P. B., 191 (262), *219*
Marshall, T. C., 120 (167), *216*
Marshall, W. D., 54 (72), 162 (216), *214, 218*
Marvel, J. T., 140 (189), *217*
Mason, W. A., 170 (231), *218*
Matsumura, F., 39 (62), 79 (103), 81 (106), 111 (157), *213, 214, 215, 216*
Maylin, G. A., 140 (188), *217*
Mehendale, H. M., 86 (117), *215*
Meikle, R. W., 117 (158), *216*
Meloan, C. E., 170 (231), *218*
Mendelson, W. L., 101 (140), *216*
Menn, J. J., 55 (78), 134 (180, 181), 136 (184), *214, 217*
Menzer, R. E., 28 (45), 29 (45), 119 (162), 156 (208), 174 (236, 237), 177 (237, 239), *213, 216, 217, 218*
Menzie, C. M., 39 (59a), *213*
Metcalf, R. L., 39 (59b, 60), 44 (67), 45 (60), 48 (67, 68), 50 (60), 51 (67, 68), 128 (174), *213, 216*
Meyer, J. A., 14 (19), 100 (138, 139), *212, 216*
Miaullis, J. B., 136 (184), *217*
Mikami, N., 72 (96, 97), 143 (191, 192), 162 (214), 163 (218), *214, 217, 218*
Mihara, K., 142 (190), 162 (214), 171 (232), *217, 218*
Miller, J. A., 101 (140), *216*
Miller, L. L., 52 (69), *213*
Minard, R. D., 20 (31), *212*
Misaki, Y., 142 (190), *217*
Misato, T., 97 (128), 106 (144), 107 (145), 133 (178), 210 (277), *215, 216, 217, 219*
Mittelstaedt, W., 102 (141), *216*
Miyamoto, J., 63 (83, 84), 71 (95), 72 (96, 97), 105 (143), 142 (190), 143 (191, 192), 162 (214), 163 (217, 218), 171 (232), 185 (247), 189 (252), *214, 216, 217*
Miyazaki, S., 39 (61), 205 (273), *213, 219*
Moilanen, K. W., 29 (43), *213*
Moldéus, P., 22 (34), *212*
Monks, I. H., 69 (94), *214*
Monselise, S. P., 140 (188), *217*
Mori, Y., 133 (178), *217*
Mumma, R. O., 91 (121, 122, 123), *215*
Mundy, D. E., 169 (227), *218*
Munger, D. M., 120 (164), *216*
Murai, T., 158 (210), 173 (234), *217, 218*
Murphy, P. J., 59 (79), *214*
Murphy, R. T., 179 (242), *218*
Murthy, N. B. K., 31 (48), *213*

N

Nakagawa, M., 33 (49), 36 (55, 56), 167 (222), *213, 218*
Nakahara, T., 127 (173), *216*
Nakamura, Y., 56 (75, 76), *214*
Nambu, K., 142 (190), *217*
Nappier, J. L., 12 (17), *212*

Neff, A. W., 8 (11), *212*
Nelson, J. O., 28, 29 (45), *213*
Newsom, H. C., 25 (37), *213*
Niki, Y., 34 (52), 56 (76), *213, 214*
Nilles, G. P., 30 (46), *213*
Nishida, T., 185 (247), *218*
Nishikawa, M., 60 (80), *214*
Nishiki, M., 97 (128), *215*
Nishiyama, K., 66 (90), *214*
Noguchi, T., 98 (129, 130, 131), *215*
Nordblom, G. D., 52 (69), *213*
North, H. H., 119 (162), *216*
Nystrom, R. F., 39 (59b, 60), 45 (60), 50 (60), *213*

O

O'Connor, M. S., 136 (185), *217*
Ogawa, K., 74 (99), 106 (144), 107 (145), *214, 216*
Ogilvie, S. Y., 23 (35), *212*
Ohkawa, H., 71 (95), 105 (143), 143 (191, 192), 163 (218), 189 (252), *214, 216, 217, 218, 219*
Okuda, I., 56 (74), *214*
Olson, L. E., 25 (38), *213*
Ost, W., 14 (20), *212*

P

Pack, D. E., 109 (150), *216*
Pagano, J. F., 101 (140), *216*
Parlar, H., 135 (183), *217*
Patil, K. C., 39 (62), 81 (106), *213, 215*
Paulson, G. D., 15 (23), 21 (33), *212*
Peake, E., 192 (254), *219*
Pease, H. L., 207 (275), *219*
Pekas, J. C., 146 (196), *217*
Petrella, V. J., 80 (104), *215*
Pickering, B. A., 154 (203), *217*
Pik, A. J., 192 (254), *219*
Pillai, C. G. P., 196 (261), *219*
Plimmer, J. R., 20 (30), 27 (40, 41), 28 (45), 29 (45), *212, 213*
Pree, D. J., 120 (166), *216*
Price, C. E., 146 (195, 196), 198 (265), *217, 219*
Pullin, E. M., 208 (276), *219*

Q

Quick, M. P., 169 (224, 227, 228), *218*

R

Ramsey, J. C., 191 (253), *219*
Rhodes, R. C., 54 (71), 67 (91), 207 (275), *214, 219*
Rice, C. P., 39 (63), *213*
Richardson, A., 81 (105), *215*
Richey, F. A., Jr., 86 (116), *215*
Ridgway, R. L., 138 (186), *217*
Robbins, J. D., 199 (267), *219*
Roberts, T. R., 6 (7), 186 (249), *212, 218*
Robinson, J., 81 (105), *215*
Rogers, H., 169 (227, 228), *218*
Rogers, R. L., 68 (92), *214*
Rose, J. Q., 191 (253), *219*
Rouchaud, J. P., 14 (19), 100 (138, 139), *212, 216*
Rowlands, D. G., 162 (215), *218*
Rueppel, M. L., 140 (189), *217*
Runnels, J., 81 (107), *215*
Rushing, D. D., 206 (274), *219*
Russel, S., 20 (31), *212*
Ruzo, L. O., 187 (250), *218*
Ryskiewich, D. P., 179 (242), *218*

S

Saeki, H., 60 (80), *214*
Safe, S., 2 (1), 11 (16), *212*
Sanga, G. K., 39 (59b), *213*
Sato, R., 31 (47), 103 (142), *213, 216*
Satogami, H., 72 (97), *214*
Sauer, H. H., 177 (240), *218*
Saunders, J. L., 120 (166), *216*
Schaefer, J., 140 (189), *217*
Scheunert, I., 20 (32), *212*
Schmidt, K., 164 (219), *218*
Schultz, D. P., 8 (10), *212*
Schultz, I. R., 98 (132), *215*
Schuphan, I., 77 (101), 125 (171), *214, 216 214, 215, 216*
Schwemmer, B. A., 132 (177), *217*
Seiber, J. N., 29 (43), *213*
Sen Gupta, A. K., 17 (25), *212*
Shaver, R., 8 (9), *212*
Shaver, R. L., 8 (12), *212*
Sheets, K. P., 86 (116), *215*
Sheets, T. J., 120 (165), *216*
Sherman, H., 99 (135), *215*
Shimabukuro, R. H., 195 (256, 257, 258, 259), 196 (257), *219*
Shimura, M., 103 (142), *216*
Shindo, N., 54 (70), *213*
Shindy, W. W., 211 (278), *219*
Shinoda, K., 97 (128), *215*

Shrivastava, S. P., 114 (156), 148 (198), *216, 217*
Sikka, H. C., 39 (63), 205 (272, 273), *213, 219*
Simon, H., 164 (219), *218*
Sims, P., 76 (100a), *214*
Singh, J., 54 (72), *214*
Slade, M., 111 (151), *216*
Smith, A. E., 13 (18), 95 (127), *212, 215*
Smith, J. N., 108 (149), *216*
Soderquist, C. J., 29 (43), 57 (77), *213, 214*
Soeda, Y., 98 (129, 130, 131), *215*
Sonawane, B. R., 118 (159), *216*
Spinnler, J. F., 62 (82), *214*
Spurr, H. W., 86 (114), *215*
Stafford, L. E., 7 (8), 15 (24), 195 (257, 258, 259), 196 (257), *212, 219*
Stallard, D. E., 85 (111, 112, 113), *215*
Standen, M. E., 186 (249), *218*
Stanley, C. W., 200 (268), *219*
Stiasni, M., 164 (219), *218*
Still, G. G., 102 (141), *216*
St. John, L. E., Jr., 43 (66), *213*
Stolzenberg, G. E., 27 (41), *213*
Stone, G. M., 13 (18), *212*
Stoutamire, D. W., 197 (264), *219*
Strang, R. H., 68 (92), *214*
Stratman, F. W., 161 (212), *217*
Strosher, M. T., 192 (254), *219*
Styrvoky, E., 41 (65), *213*
Subba-Rao, R. V., 38 (58), 42 (58), *213*
Sumida, S., 63 (83, 84), *214*
Sundström, G., 40 (64), *213*
Suss, A., 135 (183), *217*
Suzuki, H., 162 (214), *218*
Suzuki, T., 107 (146, 147, 148), 161 (213), *216, 217*
Swanson, H. R., 124 (169), 195 (256), *216, 219*
Swithenbank, C., 3 (2, 4), *212*
Szalkowski, M. B., 85 (111), *215*

T

Takahi, Y., 60 (80), *214*
Takase, I., 127 (173), 173 (235), *216, 218*
Takeda, M., 107 (146, 147, 148), *216*
Takimoto, Y., 162 (214), *218*
Tan, K. E., 173 (235), *218*
Tanaka, F. S., 7 (8), 124 (170), *212, 216*
Tashiro, S., 79 (103), *214*
Thacker, E. J., 41 (65), 149 (199), *213, 217*
Thornton, J. S., 145 (193), 200 (268), *217, 219*
Throsteinson, A. J., 39 (61), *213*
Tiedje, J. M., 9 (13), 10 (14), *212*
Tokuda, T., 97 (128), *215*
Tomida, Y., 72 (96), *214*
Tomizawa, C., 133 (179), 158 (209, 210), 173 (233, 234), *217, 218*
Tseng, C. K., 55 (78), *214*
Tsuda, M., 106 (144), 107 (145), *216*
Tsuji, H., 189 (252), *219*
Tucker, B. V., 109 (150), *216*
Tweedy, B. G., 8 (10), *212*

U

Uchiyama, M., 31 (47), 103 (142), 161 (213), *213, 216, 217*
Ueda, K., 185 (247, 248), *218*
Uesugi, Y., 133 (179), 158 (209, 210), 173, (233, 234), *217, 218*
Ueyama, I., 173 (234), *218*
Unai, T., 188 (251), *219*

V

Vadi, H., 22 (34), *212*
Van Alfen, N. K., 24 (36), *212*
VanDerSlik, A. L., 129 (175), *216*
Voss, G., 156 (207), *217*

W

Waggoner, T. B., 145 (194), *217*
Wagner, D. H., 85 (112), *215*
Wakimura, A., 163 (217), *218*
Walker, C. H., 82 (108), *215*
Wallnöfer, P. R., 2 (1), *212*
Walsh, W. C., 196 (259), *219*
Wargo, J. P., 33 (50), *213*
Wargo, J. P., Jr., 35 (54), *213*
Waring, R., 5 (6), *212*
Warrander, A., 5 (6), *212*
Watanabe, T., 103 (142), *216*
Wathana, S., 92 (124), *215*
Wayne, R. S., 18 (29), *212*
Webb, R. E., 80 (104), *215*
Weete, J. D., 196 (261), *219*
Weisgerber, I., 126 (172), *216*
Wendel, L. E., 153 (202), *217*
Whitacre, D. M., 64 (86, 87), 132 (177), *214, 217*
Whitten, C. J., 85 (110), 178 (241), *215, 218*
Wien, R. G., 124 (170), *216*
Williams, P. P., 29 (42), *213*

Williams, T. L., 59 (79), *214*
Williams, V. P., 27 (41), *213*
Wolfe, N. L., 73 (98), *214*
Wolin, R. J., 81 (107), *215*
Wong, A. S., 90 (120), 94 (125), *215*
Woo, Y.-t., 66 (90), *214*
Woodrow, J. E., 29 (43), *213*
Woods, W. G., 25 (37), *213*
Wotschokowsky, M., 14 (20), *212*

Y

Yamaguchi, I., 106 (144), 107 (165), 210 (277), *216, 219*
Yamamoto, H., 72 (96), 158 (210), *214, 217*
Yamamoto, I., 139 (187), 163 (187), *217*
Yamauchi, F., 74 (99), 106 (144), 107 (145), *214, 216*
Yang, R. S. H., 169 (225, 226), *218*
Yasuda, Y., 98 (131), *215*
Yih, R. Y., 3 (2, 4), *212*
Yllner, S., 203 (270), *219*
Yokomichi, I., 34 (52), *213*
Yoshihara, R., 63 (84), 105 (143), *214, 216*
Yost, G. A., 52 (69), *213*
Yu, C.-C., 193 (255), *219*

Z

Zabik, M. J., 30 (46), *213*
Zahlten, R. N., 161 (212), *217*
Zayed, S. M. A. D., 176 (238), *218*
Zaylskie, R. G., 15 (23), 21 (33), 27 (41), 41 (65), 124 (170), 195 (258), *212, 213, 216, 219*
Zepp, R. G., 73 (98), *214*
Zulalian, J., 18 (28, 29), 151 (201), *212, 217*

Pesticide Index

A

Abate, 166
Acid amides, 1–15
Alachlor, 9, 10
Aldicarb, 86
Aldrin, 81
Allethrin, 184
Amidines, 16–18
Anilines, 19–31
Antor, 9
Atrazine, 195–196

B

Banol, 112
Barnon, 6
BAS 305F, *see* Mebenil
BAS 3191, 11
Basalin, 30
Bassa, *see* BPMC
BATH, 129
BAY-68138, *see* Nemacur
BAY-77488, *see* Phoxim
Baygon, *see* Propoxur
BAY NYN 9306, 178
BBD, 27
Benomyl, 99–100
Benthiocarb, 56
Benzenes (substituted), 202–211
Benzoic acid 2-(2,4,6-trichlorophenyl)hydrazide, *see* BATH
Bifenox, 33, 35
Biphenyl ethers, 32–36
Bis-(*p*-chlorophenyl) acetic acid, *see* DDA
Bis-(*p*-chlorophenyl) methane, *see* DDM
Bladex, 197
BPBSMC, 113
BPMC, 107, 113
Bromophos, 164
Bromoxynil, 208
Butachlor, 9, 10
Buturon, 126
Butylate, 55
N-*sec*-Butyl-4-*tert*-butyl-2,6-dinitroaniline, *see* BBD
m-sec-Butylphenyl-*N*-methylcarbamate, *see* BPMC
m-tert-Butylphenyl-N-methylcarbamate, *see* MTBC
Bux, 109

C

Captan, 73
Carbaryl, 119
Carbofuran, 120
Carboxin, 13
cis-, trans-Chlordanes, 79
p-Chloroaniline, 20
Chloroanisidine, 23
Chlorodimeform, *see* Galecron
Chloromethoxynyl, 33, 34
Chloromethylchlor, 44
Chloroneb, 207
4-Chlorophenoxy acetic acid, *see* 4-CPA
Chloropropylate, 43
Chlorpyrifos, 168
6-Chloro-α-trichloropicoline, *see* CTP
Clearcide, 127
CNP, 33
4-CPA, 90
Credazine, 36
Cremart, 142
Crufomate, 146
CTP, 191
Cyanox, 163
Cyclophosphamide, 149
Cyolane, 150
Cypermethrins, 186
Cyprazine, 198

D

2,4-D, 91
Dasanit, 165
2,4-DB, 92
DCNA, 24
DCP, 192
DDA, 42
DDE, 40
DDM, 42

DDOD, 63
DDT, 39
o,p-DDT, 41
DDT and its analogs, 37–52
Decamethrin, 187
Diazinon, 169
Dichlorfop-methyl, 95
Dichlorobenil, 205
1,1′-(Dichloroethenylidene)bis(4-chlorobenzene), *see* DDE
Dichloronitroaniline, *see* DCNA
2,4-Dichlorophenoxyacetic acid, *see* 2,4-D
2,4-Dichlorophenoxybutyric acid, *see* 2,4-DB
3-(3,5-Dichlorophenyl)-5, 5-dimethyl-2, 4-oxazolidinedione, *see* DDOD
N-(3,5-Dichlorophenyl)succinimide, *see* DSI
3,6-Dichloro-α-picolic acid, *see* DCP
Dichlorvos, 154
Dicryl, 2
Dieldrin, 81
Diflubenzuron, *see* PH-6040
Dihydrochlordenedicarboxylic acid, 83
Dimethirimol, 69
Dimethoate, 177
Dimethylvinphos, 155
Dinitramine, 25
Dinobuton, 26
Dioxane, 66
Dioxathion, 181
Diphenamide, 8
Diphenylmethane, 38
Diphenyltrichloroethane, *see* DTE
Disugran, 206
Dithiocarbamates, 53–57
DS-15647, *see* Thiofanox
DSI, 71
DTE, 38
Dual, 9
Dyfonate, 134–135

E

Edifenfos, *see* Hinozan
Endosulfan, 78
Endrin, 80
EP-475, 118
EPTC, 55
Ethoxyaniline, 47
Ethoxychlor, 50
S-Ethyldiisobutylthiocarbamate, *see* Butylate
S-Ethyldipropylthiocarbamate, *see* EPTC
Ethylene bisdithiocarbamic acid, 54
ETU, *see* Ethylene bisdithiocarbamic acid

F

F-319, *see* Tachigaren
Fenitrothion, *see* Sumithion
Fenvalerate, 189
Flamprop-isopropyl, *see* Barnon
Fluoroimide, 74
Fonofos, *see* Dyfonate
Formothion, 177
Fthalide, 97
Furadan, *see* Carbofuran

G

Galecron, 17
GC-6506, 153
Glyphosate, 140
GS-14254, 199
Guanidines, 16–18

H

H-772, *see* Credazine
HCE, 82
Heterocyclic compounds, 58–69, 96–103
1,2,3,4,9,9-Hexachloro-*exo*-5,6-epoxy-1,4,4a,5,6,7,8,8a-octahydro-1,4-methanonaphthalene, *see* HCE
Hinozan, 173
HOE-23408(OH), *see* Dichlorfop-methyl
Hymexazol, *see* Tachigaren

I

Imides, 70–74
Inezin, 133
Ipronidazole, 61
Isothiazolinones, 62
Isoxathion, 167

K

Karphos, *see* Isoxathion
Kerb, 3
Kitazin P, 158

L

Landrin-1, 110
Landrin-2, 110, 111
Leptophos, 132
Lindane, 76

M

Machete, *see* Butachlor
Malathion, 176
MC-4379, *see* Bifenox
Mebenil, 5
Meobal, *see* MPMC
Mephosfolan, 151
Methabenzthiazuron, 102
Methamidophos, 147
Methazole, 64
Methidathion, *see* Supracide
Methiochlor, 46
Methoxychlor, 49
Methoxymethiochlor, 51
Methoxyphenone, *see* NK-049
Methylchlor, 45
Methylethoxychlor, 48
Metolachlor, *see* Dual
Mexacarbate, 117
MIPC, 106
Mipcin, *see* MIPC
MK-23, *see* Fluoroimide
Mobam, 121
Mocap, 174
Molinate, 57
Monochloroacetic acid, 203
Monolinuron, 125
Monuron, 124
MPMC, 115
MT-101, 4
MTBC, 108
MTMC, 105

N

N-2596, 136
NAC, *see* Carbaryl
1-Naphthylacetic acid, 211
Nemacur, 145
p-Nitroanisole, 22
Nitrobenzenes, 19–31
Nitrofen, 33
NK-049, 210
Norbornenes, polychlorinated, 77
Norfurazone, 68

O

OMS-1804, *see* PH-6040
Ordram, *see* Molinate
Organochlorine compounds, 75–83
Oryzemate, 103
Oxadiazone, 65
Oxamyl, 87
Oxime carbamates, 84–88
Oxyfluorofen, 33, 35

P

Parathion, 161
Parbendazole, 101
PCNB, 31
PCP, 209
Pentachloronitrobenzene, see PCNB
Pentachlorophenol, *see* PCP
Perfluidone, 15
Permethrins, 188
PH-6040, 128
Phenmedipham, 118
Phenothiol, 93
Phenoxyacetic acids, 89–95
Phenyl (aryl)carbamates, 104–121
Phenylureas, 122–130
Phosalone, 180
Phosphamidon, 156
Phosphates, 152–156
Phospholane, *see* Cyolane
Phosphonates, 137–140
Phosphonothiolates and phosphonothioates, 131–136
Phosphoramides, 144–151
Phosphoramidothiolates, 144–151
Phosphorimides, 144–151
Phosphorodithioates, 175–181
Phosphorodithiolates, 172–174
Phosphorothioamides, 141–143
Phosphorothioates, 160–171
Phosphorothiolates, 157–159
Phoxim, 170
Polychlorinated norbornenes, 77
Preforan, 33, 34
Procymidon, 72
Prometone, 198
Pronamide, *see* Kerb
Propachlor, 7
Propham, 21
Propoxur, 114
Pyrazon, 193
Pyrethrin I and II, 183

Pyrethroids, 182–189
Pyridines, 190–193
Pyrrolnitrin, 59

R

R-3828, 159
R-16661, *see* Stauffer R-16661
Resmethrins, 185
RH-315, *see* Kerb
RH-2915, *see* Oxyfluorofen
Robenidine, *see* Robenz
Robenz, 18
Roundup, *see* Glyphosate

S

S-2517, 143
Salithion, 171
SAN-6706, 68
SAN-9789, *see* Norfurazone
Silex, *see* DDOD
Sencor, 200
Sevin, *see* Carbaryl
Siduron, 123
Spartcide, *see* Fluoroimide
Stauffer R-16661, 148
Sumisclex, *see* Procymidon
Sumithion, 162
Supracide, 179
Surecide, 139

T

2,4,5-T, 94
Tachigaren, 60
2,3,6-TBA, 204
Terbacil, 67
3,4,5,6-Tetrachlorophthalide, *see* Fthalide
Tetrachlorvinphos, 155
Thiadiazuron, 130
Thiazuril, 201
Thiofanox, 85
Thiolcarbamates, 53–57
Thiophanate methyl, 98
Thioureidobenzenes, 98
TMMA, 52
Triazines, 194–201
s-Triazines, 195–199
as-Triazines, 200–201
Trichlorfon, 138
2,3,6-Trichlorobenzoic acid, *see* 2,3,6-TBA
1,1,1-Trichloro-2,2-bis(*p*-chlorophenyl) ethane, *see* DDT
N-(α-Trichloromethyl-*p*-methoxy-benzyl)-*p*-methoxyaniline
2,4,5-Trichlorophenoxyacetic acid, *see* 2,4,5-T
1,1,1-Trichlorophenyl-2-(*o*-chloro-phenyl)-2-(*p*-chlorophenyl)ethane, *see* *o,p*-DDT
Trifluralin, 28–29
Triforine, 14
Tripate, 88
Tsumacide, *see* MTMC
TTPA, 12

U

UC-34096, 116

V

Vitavax, *see* Carboxin

X

X-52, *see* Chloromethoxynyl

Z

Zectran, *see* Mexacarbate

Subject Index

A

Aedes sp., aldrin metabolism in, 81
Aedes aegypti, see Mosquitoes
Agmenellum quadraplicatum
  aldrin metabolism in, 81
  DDT metabolism in, 39
Air, trifluralin metabolism in, 29
Alfalfa
  pesticide metabolism in
    bladex, 197
    carbofuran, 120
    kerb, 3
    sencor, 200
    supracide, 179
    terbacil, 67
Alfalfa leaf cutting bee, carbaryl metabolism in, 119
Algae
  pesticide metabolism in
    aldrin, 81
    p-chloroaniline, 20
    DDT, 39
    diphenamide, 8
    ethoxyaniline, 47
    ethoxychlor, 50
    methiochlor, 46
    methoxychlor, 49
    methylchlor, 45
    PH-6040, 128
Alligator weed, dichlorobenil metabolism in, 205
Alluvial loams, clearcide metabolism in, 127
Alternalia sp.
  pesticide metabolism in
    alachlor, 10
    butachlor, 10
Alternalia mali
  pesticide metabolism in
    thiophanate methyl, 98
    thioureidobenzenes, 98
Amphidinium eartei, DDT metabolism in, 39
Anodonta, aldrin metabolism in, 81
Apis mellifera, see Bees
Apple
  pesticide metabolism in
    galecron, 17
    supracide, 179
    thiophanate, 98
    thioureidobenzenes, 98
Arachis hypogaea, see Peanuts
Asellus, aldrin metabolism in, 81
Aspergillus flavus
  pesticide metabolism in
    carbaryl, 119
    dyfonate, 134
Aspergillus fumigatus
  pesticide metabolism in
    carbaryl, 119
    dyfonate, 134
Aspergillus niger
  pesticide metabolism in
    BPMC, 107
    carbaryl, 119
    dyfonate, 134
Aspergillus terreus, carbaryl metabolism in, 119
Avena sativa, see Oats

B

Bacteria
  pesticide metabolism in
    mexacarbate, 117
    sumithion, 162
Bacteroides amylophilus, trifluralin metabolism in, 28
Bacteroides ruminicola subsp. *brevis,* trifluralin metabolism in, 28
Bacteroides succinogenes, trifluralin metabolism in, 28
Barley
  pesticide metabolism in
    atrazine, 195
    barnon, 6
    carboxin, 13
    credazine, 36
    dimethirimol, 69
    propachlor, 7
    triforine, 14
Barnes sandy loam
  pesticide metabolism in
    thiophanate methyl, 98
    thioureidobenzenes, 98

Beagles
pesticide metabolism in
benomyl, 99
dimethirimol, 69
dimethyl and tetrachlorvinphos, 155
DSI, 71
thiazuril, 201
Bean
pesticide metabolism in
carbofuran, 120
chloroneb, 207
cremart, 142
cyanox, 163
DDOD, 63
dimethoate, 177
dinobuton, 26
formothion, 177
isoxathion, 167
methazole, 64
mexacarbate, 117
mocap, 174
MTMC, 105
nemacur, 145
phosphamidon, 156
salithion, 171
sencor, 200
surecide, 139
thiophanate methyl, 98
thioureidobenzenes, 98
Bees, MTBC metabolism in, 108
Beetles, sumithion metabolism in, 162
Bermuda grass
diphenamide metabolism in, 8
phoxim metabolism in, 170
Birds
pesticide metabolism in
HCE, 82
sumithion, 162
Blood plasma
pesticide metabolism in
EP-475, 118
phenmedipham, 118
Blowfly
MTBC metabolism in, 108
Boll weevil, GC-6506 metabolite toxicity to, 153
Brevibacterium sp., 2,3,6-TBA metabolism in, 204
Broccoli
pesticide metabolism in
mexacarbate, 117
N-2596, 136
Brussels sprouts, N-2596 metabolism in, 136
Butyrivibrio fibrisolvens, trifluralin metabolism in, 28

C

Cabbage
pesticide metabolism in
isoxathion, 167
N-2596, 136
Cabbage looper, carbaryl metabolism in, 119
Canavalia ensiformis, *see* Jack bean
Carp, dinitramine metabolism in, 25
Carrot
pesticide metabolism in
atrazine, 196
cremart, 142
2,4-D, 91
Cauliflower, N-2596 metabolism in, 136
Cereal grains and straw, sencor metabolism in, 200
Chaetomium bostrychodes
pesticide metabolism in
alachlor, 10
butachlor, 10
Chaetomium globosum
pesticide metabolism in
alachlor, 10
antor, 9
butachlor, 10
dual, 9
Cherry
pesticide metabolism in
glyphosate, 140
supracide, 179
Chickens
pesticide metabolism in
aldicarb, 86
atrazine, 196
dimethyl and tetrachlorvinphos, 155
methazole, 64
perfluidone, 15
robenz, 18
thiazuril, 201
Chinese cabbage, isoxathion metabolism in, 167
Chlorella, aldrin metabolism in, 81
Chlorella pyrenoidosa, monolinuron metabolism in, 125
Chrysopa carnea, *see* Lacewing
Citrus peel, nemacur metabolism in, 145
Cladosporium cladosporioides, BPMC metabolism in, 107
Clay loam soil, chloroanisidine metabolism in, 23
Clay soil, cypermethrin metabolism in, 186
Clostridium butyricum, polychlorinated norbornenes metabolism in, 77

Cocklebur, 2,4-DB metabolism in, 92
Columba livia, see Pigeon
Compost, fthalide metabolism in, 97
Coniothyrium sp., BPMC metabolism in, 107
Corn
  pesticide metabolism in
    atrazine, 195
    bladex, 197
    butylate, 55
    2,4-D, 91
    dasanit, 165
    EPTC, 55
    mexacarbate, 117
    mocap, 174
    N-2596, 136
    PCNB, 31
    phoxim, 170
    propachlor, 7
    SAN-6706 and SAN-9789, 68
Corvus frugilegus, see Rook
Corvus monedula, see Jackdaws
Costelytra zealandica, see Grass grubs
Cotton
  pesticide metabolism in
    aldicarb, 86
    BAY NTN 9306, 178
    chloroneb, 207
    GC-6506, 153
    mephosfolan, 151
    monuron, 124
    phosphamidon, 156
    SAN-6706 and SAN-9789, 68
    Stauffer R-16661, 148
    thiofanox, 85
    trichlorfon, 138
Cotton leaf worm, malathion metabolism in, 176
Coturnix coturnix japonica, see Quail
Cow(s)
  feces, kerb metabolites in, 3
  pesticide metabolism in
    benomyl, 99
    chloropropylate, 43
    GS-14254, 199
    methazole, 64
    perfluidone, 15
    phosphamidone, 156
  urine
    kerb metabolites in, 3
Crayfish, aldrin metabolism in, 81
Cream, benomyl metabolites in, 99
Cucumber
  pesticide metabolism in
    dimethirimol, 69
    procymidon, 72
    tachigaren, 60
Cucumis melo, see Melon
Cucumis sativus, see Cucumber
Culex sp., ethoxyaniline metabolism in, 47
Culex pipiens quinquefasciatus, PH/6040 metabolism in, 128
Culex quinquefasciatus
  pesticide metabolism in
    DDT, 39
    ethoxychlor, 50
    methiochlor, 46
    methoxychlor, 49
    methylchlor, 45
Cyclops, aldrin metabolism in, 81
Cyclotella nana, DDT metabolism in, 39
Cynodon dactylon, see Bermuda grass
Cyprinus carpio, see Carp

D

Daphnia, aldrin metabolism in, 81
Daphnia magna
  pesticide metabolism in
    DDT, 39
    PH-6040, 128
    ethoxychlor, 50
    methiochlor, 46
    methoxychlor, 49
    methylchlor, 45
Daucus carota, see Carrot
Delaine ewe, nitrofen metabolism in, 33
Diatoms
  pesticide metabolism in
    aldrin, 81
    DDT, 39
Dinoflagellates, aldrin metabolism in, 81
Dog
  pesticide metabolism in
    benomyl, 99
    carbaryl, 119
    dimethirimol, 69
    dimethyl and tetrachlorinphos, 155
    DSI, 71
    thiazuril, 201
Drummer silty clay loam, glyphosate metabolism in, 140
Dugensia, aldrin metabolism in, 81
Dunaliella sp.
  pesticide metabolism in
    aldrin, 81
    DDT, 39

E

Ecosystem (model)
  pesticide metabolism in
    atrazine, 196
    chloromethylchlor, 44

DDT, 39
ethoxyaniline, 47
ethoxychlor, 50
isothiazolinones, 62
methiochlor, 46
methoxychlor, 49
methoxymethiochlor, 51
methylchlor, 45
methylethoxychlor, 48
PH-6040, 128
pyrazon, 193
Eichornis crossiper, see Water hyacinth
Environment, *see* Ecosystem
Escherichia coli B, DCNA metabolism in, 24
Estigmene acrea, see Saltmarsh caterpillar
Eubacterium limosum, trifluralin metabolism in, 28
Eubacterium ruminanticum, trifluralin metabolism in, 28
Euonymous alatus, diphenamide metabolism in, 8

F

Fargo silty loam
pesticide metabolism in
thiophanate methyl, 98
thioureidobenzenes, 98
Feces
aldrin metabolites in, 81
benomyl metabolites in, 99
cyclophosphamide metabolites in, 149
Fish
pesticide metabolism in
DDT, 39
ethoxyaniline, 47
ethoxychlor, 50
methiochlor, 46
methylchlor, 45
methoxychlor, 49
PH-6040, 128
Fluorescent lamp, nitrofen and CNP metabolism by, 33
Fungi
pesticide metabolism in
sumithion, 162
thiophanate methyl, 98
thioureidobenzenes, 98
Fusarium oxysporum
pesticide metabolism in
BPMC, 107
carbaryl, 119
p-chloroaniline, 20
dyfonate, 134
Fusarium roseum
pesticide metabolism in
alachlor, 10
butachlor, 10
carbaryl, 119

G

Gambusia affinis, dyfonate toxicity to, 134
Gambusia affinis
pesticide metabolism in
DDT, 39
ethoxyaniline, 47
ethoxychlor, 50
methiochlor, 46
methoxychlor, 49
methylchlor, 45
PH-6040, 128
Gammarus, aldrin metabolism in, 81
Geotrichum candidum, carbaryl metabolism in, 119
Germicidal lamp, methazole metabolism by, 64
Glass, dioxathion metabolism on, 181
Gliocladium roseum, carbaryl metabolism in, 119
Gloecystis sp., diphenamide metabolism in, 8
Glycine max, see Soybean
Goat
pesticide metabolism in
GS-14254, 199
phosphamidon, 156
propham, 21
Gossypium hirsuntum, see Cotton
Grape
pesticide metabolism in
DDOD, 63
supracide, 179
thiophanate methyl, 98
thioureidobenzenes, 98
Grass, dasanit metabolism in, 165
Grass grubs, MTBC metabolism in, 108
Guinea pig
pesticide metabolism in
carbaryl, 119
mebenil, 5

H

Hagerstown soil, kerb metabolism in, 3
Halobdella stangalis, aldrin metabolism in, 81
Helianthus annus, see Sunflower
Heliothis virescens, see Tobacco bud-worm
Helminthosporium sp., carbaryl metabolism in, 119
Hops, supracide metabolism in, 179
Hordeum vulgare, see Barley

Housefly
dyfonate toxicity to, 134
pesticide metabolism in
abate, 166
DDT, 39
diazinon, 169
dinobuton, 26
ethoxyaniline, 47
ethonychlor, 50
landrin, 111
methamidophos, 147
methiochlor, 46
methoxychlor, 49
methylchlor, 45
mexacarbate, 117
MTBC, 108
MTMC, 105
propoxur, 114
Stauffer R-16661, 148
Human
carbaryl metabolism in, 119
feces, aldrin metabolism in, 81
Human embryonic lung cells, carbaryl metabolism in, 119
Hydra, aldrin metabolism in, 81
Hydrogenomonas sp.
pesticide metabolism in
DDA, 42
DDM, 42
diphenylmethane, 38
DTE, 38
Hydrolysis
pesticide metabolism in
captan, 73
ethylene bisdithiocarbamic acid, 54
isothiazolinones, 62
triforine, 14

I

Insects, *see* individual types
Inu-apple tree, fluoroimide metabolism in, 74
Iowa silt soil, BUX metabolism in, 109

J

Jack bean, 2,4-D metabolism in, 91
Jackdaw, HCE metabolism in, 82
Jameson sandy loam, dichlorfop-methyl metabolism in, 95

K

Keyport silt loam, chloroneb metabolism in, 207
Kinnow mandarin fruiting branch, 1-naphthylacetic acid metabolism in, 211
Kochi mineral soils
pesticide metabolism in
BPMC, 107
MIPC, 106
Kumagaya clay, fluoroimide metabolism in, 74

L

Lacewing, trichlorofon metabolism in, 138
Lachnospira multiparus, trifluralin metabolism in, 28
Lactieca sativa, see Lettuce
Lakeland fine sand, aldicarb metabolism in, 86
Leech, aldrin metabolism in, 81
Lettuce, atrazine metabolism in, 196
Light
pesticide metabolism by
PH-6040, 128
surecide, 139
Litonia sandy loam, glyphosate metabolism in, 140
Liver
pesticide metabolism in
ethoxyaniline, 47
ethoxychlor, 50
methylchlor, 45
Loam soil
pesticide metabolism in
cyanox, 163
DCP, 192
kerb, 3
surecide, 139
Loess soil, methabenzthiazuron metabolism in, 102
Lucia scricata, see Blowfly
Lufkin fine sandy loam
pesticide metabolism in
aldicarb, 86
BAY NTN 9306, 178
Lung cell culture
pesticide metabolism by
banol, 112
UC-34096, 116
Luvisol, DCP metabolism in, 192
Lycopersicon esculentum, see Tomato
Lygus bugs, trichlorfon metabolism in, 138
Lygnus hesperus, see Lygnus bugs
Lymnaea, aldrin metabolism in, 81

M

Malus pruniforia, see Inu-apple tree
Malus pumila, see Apple
Mandarin fruiting branch, 1-naphthylacetic acid metabolism in, 211
Marsh grass, atrazine metabolism in, 196
Meal worms, MTBC metabolism in, 108
Megachile pacifica, see Alfalfa leaf cutting bee
Melfort silty clay, dichlorfop-methyl metabolism in, 95
Melon, benomyl metabolism in, 100
Mercury vapor lamp
  pesticide metabolism in
    BBD, 27
    benthiocarb, 56
    cyanox, 163
Merion Kentucky bluegrass, siduron metabolism in, 123
Metapeake loam soil
  pesticide metabolism in
    oxadiazone, 65
    phosalone, 180
Microsomes
  pesticide metabolism in
    aldrin, 81
    benomyl, 100
    BPBSMC and BPMC, 113
    carbaryl, 119
    crufomate, 146
    diazinon, 169
    dihydrochlordenedicarboxylic acid, 83
    dimethoate, 177
    dimethyl and tetrachlorvinphos, 155
    dinobuton, 26
    dioxathion, 181
    dyfonate, 134
    endrin, 80
    EP-475, 118
    ethoxylaniline, 47
    ethoxychlor, 50
    ethylene bisdithiocarbamic acid, 54
    formothion, 177
    galecron, 17
    HCE, 82
    ipronidazole, 61
    landrin, 111
    methamidophos, 147
    methylchlor, 45
    mexacarbate, 117
    mocap, 174
    MTBC, 108
    MTMC, 105
    nemacur, 145
    p-nitroanisole, 22
    oxamyl, 87
    parathion, 161
    PH-6040, 128
    phenmedipham, 118
    phosphamidon, 156
    propoxur, 114
    pyrrolnitrin, 59
    S-2517, 143
    Stauffer R-16661, 148
    trifluralin, 29
Microtus pitymus pinetorum, see Pine mouse
Milk
  benomyl metabolites in, 99
  dasanit metabolites in, 165
Millet, chloromethoxynyl metabolism in, 34
Monkey
  pesticide metabolism in
    carbaryl, 119
    thiazuril, 201
Monmouth fine sandy loam soil
  pesticide metabolism in
    oxadiazone, 65
    phosalone, 180
Montcalm sandy loam
  basalin metabolism in, 30
Mosquitoes
  pesticide metabolism in
    abate, 166
    DDT, 39
    ethoxyaniline, 47
    ethoxychlor, 50
    methiochlor, 46
    methoxychlor, 49
    methylchlor, 45
    PH-6040, 128
    Stauffer R-16661, 148
Mouse
  pesticide metabolism in
    benomyl, 99
    benthiocarb, 56
    chloromethylchlor, 44
    DDT, 39
    dichlorvos, 154
    dinobuton, 26
    ethoxychlor, 50
    galecron, 17
    landrin, 111
    methamidophos, 147
    methiochlor, 46
    methoxychlor, 49
    methoxymethiochlor, 51
    methylchlor, 45
    methylethoxychlor, 48

monochloroacetic acid, 203
MPMC, 115
MTBC, 108
phosphamidon, 156
thiophanate methyl, 98
thioureidobenzenes, 98
Mucor sp.
pesticide metabolism in
BAS 3191, 11
carbaryl, 119
Mucor altenans, dyfonate metabolism in, 134
Mucor plumbeus, dyfonate metabolism in, 135
Mucor racemosus, carbaryl metabolism in, 119
Mugho pine, carbofuran metabolism in, 120
Musca domestica, see Housefly
Mussels, aldrin metabolism in, 81
Myriophyllum brasilience, see Parrot feather

N

Norfolk sandy loam
pesticide metabolism in
aldicarb, 86
glyphosate, 140
Nicotiana tabacum, see Tobacco
Nitzschia sp. DDT metabolism in, 39

O

Oats, atrazine metabolism in, 195–196
Oedogonium sp
pesticide metabolism in
diphenamide, 8
ethoxyaniline, 47
Oedogonium cardiacum
pesticide metabolism in
DDT, 39
ethoxychlor, 50
methiochlor, 46
methoxychlor, 49
methylchlor, 45
PH-6040, 128
Olisthodiscus luteus, DDT metabolism in, 39
Oryza sativa, see Rice

P

Paddy field soils
pesticide metabolism in
clearcide, 127
MIPC, 106
Paecilomyces sp.
pesticide metabolism in
alachlor, 10
BBD, 27
butachlor, 10
Panicum crus-galli var. *frumentaceum, see* Millet
Paracoccus sp., p-chloroanitine metabolism in, 20
Parrot feather
pesticide metabolism in
dichlorobenil, 205
diphenamide, 8
Peach orchard soil, atrazine metabolism in, 196
Peanuts
pesticide metabolism in
nemacur, 145
perfluidone, 15
Peas, atrazine metabolism in, 195
Pellicularia sasakii
pesticide metabolism in
thiophanate methyl, 98
thioureidobenzenes, 98
Penicillium sp.
pesticide metabolism in
alachlor, 10
butachlor, 10
carbaryl, 119
Penicillium funiculosum, BPMC metabolism in, 107
Penicillium notatum, dyfonate metabolism in, 135
Penicillium roqueforti, carbaryl metabolism in, 119
Pepstostreptococcus elsdenii, trifluralin metabolism in, 28
Phaseolus vulgaris, see Bean; Snapbean
Phoma sp.
pesticide metabolism in
alachlor, 10
butachlor, 10
Photolysis
pesticide metabolism by
benomyl, 100
isothiazolinones, 62
monuron, 124
sumithion, 162
TMMA, 52
trifluralin, 29
Physa sp.
pesticide metabolism in
DDT, 39

ethoxyaniline, 47
ethoxychlor, 50
methiochlor, 46
methoxychlor, 49
methylchlor, 45
PH-6040, 128
Pig, carbaryl metabolism in, 119
Pigeon, HCE metabolism in, 82
Pine mouse, endrin metabolism in, 80
Pineapple, nemacur metabolism in, 145
Pinus mugo, see Mugho pine
Pisum, sativum, see Peas
Poa pratensis, see Merion Kentucky bluegrass
Pond water and sediment, dichlorobenil metabolism in, 205
Potamogeton diversifolius, see Waterthread
Potatoes
pesticide metabolism in
dyfonate, 134
nemacur, 145
sencor, 200
supracide, 179
thiofanox, 85
Protozoa, aldrin metabolism in, 81
Prunes, supracide metabolism in, 179
Pseudomonas cepacia, DCNA metabolism in, 24
Pseudomonas putida
pesticide metabolism in
DDA, 42
DDM, 42
diphenylmethane, 38
DTE, 38
Pyricularia oryzae
pesticide metabolism in
hinozan, 173
inezin, 133
kitazin P, 158

Q

Quail, HCE metabolism in, 82

R

Rabbit
pesticide metabolism in
benomyl, 99
formothion, 177
HCE, 82
mebenil, 5
mocap, 174
S-2517, 143
Rat(s)
pesticide metabolites in
isothiazolinones, 62
kerb, 3
liver
dihydrochlordenedicarboxylic acid, 83
dinobuton metabolism in, 26
galecron metabolism in, 17
p-nitroanisole metabolism in, 22
oxamyl, 87
parathion, 161
trifluralin metabolism in, 29
pesticide metabolism in
allethrin, 184
banol, 112
benomyl, 99
bladex, 197
BPBSMC and BPMC, 113
butylate, 55
carbofuran, 120
chlordanes, 79
credazine, 36
cremart, 142
crufomate, 146
CTP, 191
cyanox, 163
cyolane, 150
cyprazine, 198
DDE, 40
DDOD, 63
o,p'-DDT, 41
diazinon, 169
dichlorvos, 154
dimethirimol, 69
dimethoate, 177
dimethyl and tetrachlorinphos, 155
dinobuton, 26
dioxane, 66
dioxathion, 181
DSI, 71
dyfonate, 134
EP-475, 118
EPTC, 55
fenvalerate, 189
formothion, 177
HCE, 82
leptophos, 132
lindane, 76
mebenil, 5
methazole, 64
mobam, 121
mocap, 174
molinate, 57

MPMC, 115
MTMC, 105
N-2596, 136
NK-049, 210
oxyfluorofen, 35
perfluidone, 15
permethrins, 188
phenmedipham, 118
phosphamidon, 156
procymidon, 72
prometone, 198
propham, 21
pyrethrins, 183
pyrrolnitrin, 59
resmethrins, 185
robenz, 18
salithion, 171
thiadiazuron, 130
thiazuril, 201
trichlorfon, 138
triforine, 14
TTPA, 12
urine
pesticide metabolites in
kerb, 3
prometone, 198
Ray silt loam, glyphosate metabolism in, 140
Regina heavy clay, dichlorfop-methyl metabolism in, 95
Reticulitermis flavipes, see Termites
Rhizopus sp., carbaryl metabolism in, 119
Rhizopus arrhizus, dyfonate metabolism in, 135
Rhizopus japonicus
pesticide metabolism in
BAS 3191, 11
dicryl, 2
dyfonate, 135
Rhizopus nigricans, BAS 3191 metabolism in, 11
Rhizopus peka, BAS 3191 metabolism in, 11
Rice
pesticide metabolism in
BPMC, 107
chloromethoxynyl, 34
cremart, 142
2,4-D, 91
hinozan, 173
inezin, 133
kitazin P, 158
MIPC, 106
nitrofen, 33
oryzemate, 103
oxadiazone, 65
salithion, 171
tachigaren, 60
River water system, isothiazolinone metabolism in, 62
Rook, HCE metabolism in, 82
Rumen fluid
pesticide metabolism in
benomyl, 100
disugran, 206
Rumen microorganisms, triflularin metabolism in, 28
Ruminicoccus flavefaciens, trifluralin metabolism in, 28

S

Saccharum officianarum, see Sugarcane
Sacramento clay soil, molinate metabolism in, 57
Saltmarsh caterpillar
pesticide metabolism in
chloromethylchlor, 44
DDT, 39
ethoxyaniline, 47
ethoxychlor, 50
methiochlor, 46
methoxychlor, 49
methoxymethiochlor, 51
methychlor, 45
methylethoxychlor, 48
PH-6040, 128
Sandy clay soil, cypermethrins metabolism in, 186
Sandy loam soil
pesticide metabolism in
aldicarb, 86
cypermethrins, 186
DCP, 192
Scenedesmus sp., diphenamide metabolism in, 8
Setagaya top soil
pesticide metabolism in
cyanox, 163
surecide, 139
Sheep
pesticide metabolism in
benomyl, 99
carbaryl, 119
cyclophosphamide, 149
diazinon, 169
disugran, 206
nitrofen, 33

parbendazole, 101
PH-6040, 128
thiophanate methyl, 98
thioureidobenzenes, 98
Silica gel
pesticide metabolites in
basalin, 30
dioxathion, 181
Skeletonema costatum, DDT metabolism in, 39
Sludge (activated), isothiazolinone metabolism in, 62
Snails
pesticide metabolism in
aldrin, 81
DDT, 39
ethoxyaniline, 47
ethoxychlor, 50
methiochlor, 46
methoxychlor, 49
methylchlor, 45
Snapbean
pesticide metabolism in
chloroneb, 207
dioxathion, 181
ethylene bisthiocarbamic acid, 54
landrin, 111
Soil(s)
pesticide metabolism in
aldicarb, 86
BAY NTN 9306, 178
benthiocarb, 56
buturon, 126
BUX, 109
chloroneb, 207
CNP, 33
cyanox, 163
cypermethrins, 186
DCP, 192
dichlorfop-methyl, 95
ethylene bisdithiocarbamic acid, 54
methabenzthiazuron, 102
MT-101, 4
nitrofen, 33
PCNB, 31
PCP, 209
PH-6040, 128
phosalone, 180
procymidon, 72
sumithion, 162
supracide, 179
surecide, 139
thiofanox, 85
Solanum tuberosum, see Potatoes
Sorghum
pesticide metabolism in
atrazine, 195
DDT, 39
ethoxychlor, 50
methiochlor, 46
methoxychlor, 49
methylchlor, 45
PH-6040, 128
propachlor, 7
Sorghum bicolor, see Sorghum
Sorghum halpense, see Sorghum
Sorghum vulgare, see Sorghum
Soybean
pesticide metabolism in
atrazine, 196
2,4-D, 91
2,4-DB, 92
diphenamide, 8
mexacarbate, 117
SAN-6706 and SAN-9789, 68
Spartina alterniflora, atrazine metabolism in, 196
Spider mites, galecron metabolism in, 17
Spinach, parathion metabolism in, 161
Spodoptera littoralis, see Cotton leaf worm
Steer, R-3828 metabolism in, 159
Streptococcus bovis, trifluralin metabolism in, 28
Succinimonas amylolytica, trifluralin metabolism in, 29
Succinivibrio dextrinosolvens, trifluralin metabolism in, 29
Sugarbeets
pesticide metabolism in
N-2596, 136
thiofanox, 85
Sugarcane
pesticide metabolism in
atrazine, 195
propachlor, 7
sencor, 200
Summer wheat, dichlorfop-methyl metabolism in, 95
Sunflower, 2,4-D metabolism in, 91
Sunlight
pesticide metabolism by
BBD, 27
benomyl, 100
BPBSMC and BPMC, 113
CNP, 33
decamethrin, 187

dinitramine, 25
dinobuton, 26
landrin, 111
methazole, 64
nitrofen, 33
2,4,5-T, 94
triforine, 14

T

Tenebrio sp., *see* Meal worms
Termites, chlorpyrifos metabolism in, 168
Tetranychus urticae, see Spider mites
Tetraselmis chuii, DDT metabolism in, 39
Tobacco
pesticide metabolism in
2,4-D, 91
carbaryl, 119
carbofuran, 120
endosulfan, 78
nemacur, 145
tripate, 88
Tobacco bud-worm, trichlorfon metabolism in, 138
Tochigi clay loam soils
pesticide metabolism in
BPMC, 107
fluoroimide, 74
MIPC, 106
Tomatoes
pesticide metabolism in
bromophos, 164
credazine, 36
diphenamide, 8
ethylene bisdithiocarbamic acid, 54
nemacur, 145
sencor, 200
tachigaren, 60
Towner loamy find sand
pesticide metabolism in
thiophanate methyl, 98
thioureidobenzenes, 98
Trichoplusia ni, see Cabbage looper
Tribolium castaneum, see Beetles
Trichoderma sp.
pesticide metabolism in
alachlor, 10
butachlor, 10
Trichoderma viride
pesticide metabolism in
carbaryl, 119
dyfonate, 135
mexacarbate, 117
Triticum aestivum, see Summer wheat or Wheat
Triticum vulgare, see Wheat
Turkey, iprondazole metabolism in, 61

U

Udenfriend model system, Stauffer R-16661 metabolism in, 148
Ulothrix sp., diphenamide metabolism in, 8
Urine
pesticide metabolites in
benomyl, 99
cyclophosphamide, 149
isothiazolinones, 62
kerb, 3
prometone, 198
UV light
pesticide metabolism in
alachlor, 10
BATH, 129
BPBSMC and BPMC, 113
butachlor, 10
4-CPA, 90
decamethrin, 187
diazinon, 169
dinitramine, 25
ethylene bisdithiocarbamic acid, 54
S-2517, 143
2,4,5-T, 94
triforine, 14

V

Vitis viniferalabrusca, see Grape
Volcanic ash silty loam, clearcide metabolism in, 127

W

Water
pesticide metabolism in
buturon, 126
p-chloroaniline, 20
glyphosate, 140
2,4,5-T, 94
thiofanox, 85
UC-34096, 116

Water hyacinth, diphenamide metabolism in, 8
Waterthread, diphenamide metabolism in, 8
Wheat
  pesticide metabolism in
    atrazine, 196
    bromoxynil, 208
    carboxin, 13
    diphenamide, 8
    nitrofen, 33
    phoxim, 170

X

Xanthium sp., *see* Cocklebur
XD cell line
  pesticide metabolism in
    carbaryl, 119
    preforan, 34

Z

Zea mays, see Corn